D0758245

Wattana

Wattana

AN ORANGUTAN IN PARIS

Chris Herzfeld

Translated by Oliver Y. Martin and Robert D. Martin

The University of Chicago Press

Chicago and London

Chris Herzfeld is a philosopher of science and an artist. She is a founder of the Great Apes Enrichment Project, and the author or coauthor of two other books on primates. She divides her time between Paris, Brussels, and Naples, Florida. **Oliver Y. Martin** is part of the Institute for Integrative Biology at ETH Zurich. **Robert D. Martin** is curator emeritus at the Field Museum, Chicago, and the author of *How We Do It: The Evolution and Future of Human Reproduction*.

The University of Chicago Press, Chicago 60637
The University of Chicago Press, Ltd., London

Printed in the United States of America

25 24 23 22 21 20 19 18 17 16 1 2 3 4 5

ISBN-13: 978-0-226-16859-3 (cloth)
ISBN-13: 978-0-226-16862-3 (e-book)
DOI: 10.7208/chicago/9780226168623.001.0001

Originally appeared in French as *Wattana: Un orang-outan à Paris*, © Editions Payot et Rivages, 2012

Publication of this book has been aided by a grant from CNL.

Library of Congress Cataloging-in-Publication Data
Herzfeld, Chris, author.
[Wattana, un orang-outan à Paris. English]
Wattana, an orangutan in Paris / Chris Herzfeld ; translated by Oliver Y. Martin and Robert D. Martin.
pages cm
"Originally appeared in French as Wattana : un orang-outan à Paris, © Editions Payot et Rivages, 2012."—Title page verso.
Includes bibliographical references.
ISBN 978-0-226-16859-3 (cloth : alkaline paper)—ISBN 978-0-226-16862-3 (e-book) 1. Wattana (Orangutan) 2. Orangutans—Behavior. 3. Captive wild animals—Behavior. I. Title.
QL737.P96H4613 2016
599.88'315—dc23 2015027960

♾ This paper meets the requirements of ANSI/NISO Z39.48-1992 (Permanence of Paper).

CONTENTS

INTRODUCTION

The animal! Somber mystery! . . . Immense world of mute dreams and pains. . . . Away with prejudice, and look at their mild and dreamy air, and the attraction which the most advanced among them evidently feel for man.

—JULES MICHELET, *The People*, 126.

In the heart of the Gombe Forest, in Tanzania, Jane Goodall gradually succeeded in becoming accepted by the chimpanzees as a peripheral member of their troop. This closeness, the links she was able to forge with them, and the long years spent in the field from 1960, allowed Jane to tackle a hitherto poorly known dimension in the life of great apes: She revealed personality traits, explored their histories, and established their genealogies. At the beginning of the monograph she dedicated to them, she hence provides portraits of several chimpanzees, each portrait illustrated with the individual's photograph and its name.[1] A few years after Goodall had set up her research site in Tanzania, Dian Fossey founded her study centre in 1966 in Kabara, in the former Belgian Congo. She went on to share daily life with the mountain gorillas for eighteen years. In 1983, she too shared their characters and biographies with readers of her bestseller *Gorillas in the Mist.*[2] By becoming so keenly interested in the personalities of the apes they observed, these two primatologists thus went beyond one of the fundamental directives of science, which focuses not on individuals, but on species and populations as objects of study. There is, after all, no science without generalizations.

Captive Great Apes

Nevertheless, even if some field primatologists have been able to tell the stories of the apes they have met in the wild, few authors have shown interest in the thousands of great apes[3] that reside in our zoos. Of course, this does not mean that these captive chimpanzees, bonobos, gorillas, and orangutans do not have a life, a history, an existence. Yet, they remain anonymous, moved from zoo to zoo and replaced in their cages, until—inexorably—they disappear. Only the zookeepers, the veterinarians, and a few faithful visitors take interest in their destinies and would be able to reconstruct their lineages. These individual apes leave almost no trace of their existence beyond being mentioned in a *Studbook*[4] or, in less coldly administrative terms, in anecdotes such as those recounted by naturalists, zoo directors,[5] biologists, or specialized writers.[6] This book tries to play the role of spokesperson for these somewhat forgotten primates, by focusing on the story of a female orangutan who lived for several years in the Ménagerie of the Jardin des Plantes: Wattana. Through her, we will explore the world of captive great apes, and study the fate of our cousins living in close proximity among humans. Called as a witness, Wattana will help us understand the everyday life of the great apes, the links that are forged between them and their keepers, the shock of being transferred from zoo to zoo, the lives and deaths of these apes who show so much goodwill when obliged to be part of our world, yet whose existence is mainly met with indifference rather than real empathy. She will also allow me to evoke multiple, interrelated histories, like so many arenas where orangutans will take pride of place: the history of encounters between human beings and great apes, that of the

Ménagerie of the Jardin des Plantes, that of captive orangutans, and that of the great apes who have lived in the Ménagerie.

Orangutans

Wattana is a Bornean orangutan. Among great apes,[7] the two species of orangutans, natives of Borneo and Sumatra, are the only ones to live on the Asian continent.[8] The others—bonobos, chimpanzees, and gorillas—are all native to Africa. As a group, these species, together with human beings, belong to the superfamily Hominoidea.[9] As is well known, humans are not *descended* from apes, but instead share ancestors with them. Humans are part of the same phylogenetic group[10] as the other hominoid primates. The *last common ancestors* of chimpanzees and humans are thought to have lived six million years ago. The genetic difference between humans and gorillas is greater than that between humans and chimpanzees (or bonobos). An orangutan shares approximately 97% of its genes with humans, and it is the great ape that is the least closely related to our species. Orangutans are thought to be related to populations of hominoids that probably colonized certain parts of Asia,[11] several million years ago. This migration was rendered possible by the existence of land bridges, now covered by seas. In earlier classifications, the orangutans of Borneo (*Pongo pygmaeus,* Linné 1760) and those of Sumatra (*Pongo abelii,* Lesson 1827) were ranked as different subspecies. Nowadays, they are generally accepted as distinct species in their own right, thought by some to have diverged from each other about 400,000 years ago.[12] The orangutans of Sumatra are characterized by their slenderness, the presence of beards, hair that is tinged with yellow on the lower part of their faces, the golden brown color of their eyes, as well as longer and silkier bright orange fur. They are also thought to be more social and arboreal.[13] As far as reproduction is concerned, it seems that there are far fewer incidences of forced mating in Sumatran orangutans (50%), whereas coercive copulations are very frequent in Bornean orangutans, occurring in 90% of sexual encounters. The Bornean species includes three subspecies: *Pongo pygmaeus pygmaeus* (in Sarawak, Malaysia, and in the northwestern province of Kalimantan,[14] Borneo), *Pongo pygmaeus wurmbii* (Central Kalimantan), and *Pongo pygmaeus morio*[15] (Sabah region in Malaysia and Eastern Kalimantan).[16] Between 45,000 and 69,000 orangutans live on Borneo[17] (threatened with extinction), while 7,300 orangutans inhabit Sumatra[18] (critically endangered). Orangutans are threatened by human activities that lead to fragmentation and loss of habitat, caused by converting forests into agricultural zones (in order to prioritize particular monocultures, mainly that of palm oil), mining operations, as well as construction of the roads needed for these different kinds of commercial enterprise. Illegal trade in wood and rare species, forest fires, and hunting pose additional threats. As a result of widespread corruption, Indonesia has lost 80% of its forest cover, with its nature reserves also affected. Some experts believe that orangutans will have disappeared from the wild by 2030. The fateful encounters between these Asian apes and human beings have led to a dramatic reduction in their populations, resulting in the loss of some 60% of members of the Bornean species over the last sixty years and of about 80% of members of the Sumatran species during the past seventy-five years. Both species are rendered even more vulnerable by decreased rates of genetic admixture because of habitat fragmentation, combined with long intervals between births: between four and nine years depending on the species, region, and individual.

A Long Shared History

Contacts between humans and great apes are not a recent development, but began in Antiquity. In the fifth or sixth century BC, gorillas were mentioned by the Carthaginian admiral Hanno, who observed several specimens during a voyage of exploration along the coast of West Africa. Hanno tried to bring back representatives of this unknown species, but his soldiers, attacked and bitten by the three females they attempted to capture, were forced to kill the apes. Later, in the second century BC, Galen dissected apes, describing their anatomy as if they were human, the study of human corpses being in fact forbidden in Rome at the time. Much later, in the sixteenth century, Andreas Vesalius, suspecting the substitution of apes for humans, organized a public dissection in Bologna. He was able to confirm his hypothesis by studying the anatomy of a human in comparison to a great ape. From the fifteenth century on, many ambitious voyages of exploration were funded by a few powerful Western nations, and participating navigators, soldiers, missionaries, or doctors mentioned pongos (orangutans) and jockos (chimpanzees) in their accounts. Great apes arrived in Europe in the seventeenth century. Nicolaes Tulp described a chimpanzee sent from Angola to the Prince of Orange, the ape having arrived in good shape in the Netherlands. Nevertheless, most primates were studied as corpses, as they rarely survived their capture and subsequent transport. In the eighteenth century, studies were at last carried out on live specimens. The great French naturalist of the eighteenth century Buffon welcomed a chimpanzee, the famous Jocko, in his home. He relates how the chimp displayed refined manners when drinking tea. The Dutch scientist Arnout Vosmaer describes the behavior of an orangutan kept in one of the menageries of the Prince of Orange. He recounts how the ape had adopted several human habits. At the beginning of the next century, the first great apes began to appear as exhibits in zoos. During the colonial expansions, they started arriving in large numbers in the West. They were then consigned to zoos and laboratories, where they were roped into medical experiments or anatomical studies, or used as study objects in the fields of ethology or comparative psychology. In the twentieth century, research has focused on exploring their intelligence, investigating their language capacities, and analyzing their genetic heritage, while their use in biomedical research has expanded because of their close relatedness to humans. Following a few pilot studies in the wild in the 1960s and 1970s, several long-term field studies were initiated by primatologists, both Japanese (Imanishi, Itani, Nishida, Kano) and Anglo-Saxon (Goodall, Fossey, Galdikas).

Existence in a Zoo

Great apes that spend their entire lives in zoos are rarely mentioned in the history of our relationship with primates. The chimpanzees of Gombe forest described by Jane Goodall, or the bonobos of the faraway forest regions of the Congo studied by Takayoshi Kano, exert more appeal, cloaked as they are in authenticity and exoticism. The captive great apes that the public is so keen to see in zoos frequently amuse zoo visitors with their facial gestures and imitations. But these visitors know almost nothing about the objects of their amusement, and captive apes are often reduced to being animals destined to entertain us on a Sunday afternoon, supposed witnesses of our distant origins, and representatives of a natural world from which city dwellers are now isolated. During their brief tours of the sections reserved for primates, visitors lack the time to become interested in the character traits specific to each individual, or to enter their world, even for a short time. Yet, if one does stop and pay attention, the residents of these places of encapsulated exoticism are truly fascinating. Forced to make a life in our worlds, they show incredible creativity, and wander alone with responsive ease among our objects, our habits, our techniques. They accordingly adopt our life skills, knowledge, and capacities, and transform these according to their needs and on their own terms. They then exhibit surprising behaviors and display a plasticity that is unexpected from creatures too quickly ranked among animals that are environmentally depauperate, existing as *Nur-Lebenden* (merely alive) in a pure and simple state, just living and nothing more, as described by Heidegger.[19] Among their capacities to adopt various human skills, our awareness generally centers on the extraordinary aptitude for imitation that great apes show. Yet, these capacities extend much further. For example, Wattana has learned how to tie knots. She has even acquired a true *mastery* of tying knots, using fine manipulative skills long thought to be exclusive to humans. Another orangutan, Nénette, has occasionally been observed quite deliberately cleaning the glass panes of her enclosure with a rag. The gorilla Victoria has also been observed tracing drawings on the windows of her cage. The bonobo Hermien has acquired the habit of smiling to greet his human companions, despite the fact that showing the teeth is generally connected with threatening expression in primates. Furthermore, when performing some of these activities borrowed from humans, great apes display a remarkable sense of aesthetics, evident enjoyment, and even *Funktionslust,* a phenomenological concept that expresses the pleasure felt when doing what one knows how to do well. These behaviors are by no means remote-controlled in ways that could be explained by training, conditioning, or mechanical imitation of gestures. Living in close proximity to humans, great apes integrate the human universe to a degree that few had thought possible, going so far as to adopt their ethos, their way of existing on earth. Wattana undoubtedly counts among those apes that are fascinated by the possibilities on offer at the heart of our cultures.

1 The Ménagerie of the Jardin des Plantes

What is it that impresses me in an animal? The first thing that impresses me is the fact that every animal has a world.
—GILLES DELEUZE, "A for Animal," trans. Charles J. Stivale

In the aftermath of the French Revolution, after the departure of the royal family from the Château de Versailles, the menagerie of Louis XVI, King of France and Navarre, had been ransacked before the eyes of the concierge, Sieur Laimant, and condemned to destruction. The Regisseur of the Estate of Versailles, Louis Charles Couturier, wrote to the naturalist Jacques-Henri Bernardin de Saint-Pierre, who went to Versailles accompanied by two botanists of the Muséum, René Louiche Desfontaines and Jean-André Thouin. The surviving inhabitants of the Royal Ménagerie were promised to Bernardin de Saint-Pierre, with the expectation that they would end up as stuffed specimens for exhibition in the museum.[1] But the naturalist wanted to keep the animals alive. So he planned to open an animal park in the French capital[2]—a park destined to delight not just the merely curious but also scientists, illustrators, painters, and poets alike. Earlier, Louis Jean-Marie

Daubenton, professor of anatomy who had contributed to the famous Buffon's *Histoire naturelle des Animaux*, had already argued for the same project. On December 14, 1792, a report written by three professors of the Muséum to support the creation of a menagerie was presented at the Natural History Society of Paris (founded in 1787 as the *Société linnéenne de Paris*).[3] Yet it was not the professors of the Jardin des Plantes who were responsible for creating a zoo in Paris, but rather an unexpected development that took the museum's members by surprise. At the time, showmen, merchants, and animal exhibitors would frequently display wild animals on the city's streets or plazas. The general council of the municipality of Paris saw this as a danger to public safety, and such animals were confiscated from November 1793 onward. After their removal from the streets, these animals had to be housed somewhere, so it was decreed that they should be taken to the Jardin des Plantes. Thus it was that the first national menagerie in Europe[4] came into being as a byproduct of the revolution. The reborn Ménagerie was opened to the public, in deliberate contrast with the royal menagerie, an institutional symbol of the pride of tyrants accessible only to the privileged. The few remaining animals displayed in the royal menagerie[5] were requisitioned and transported to Paris. On April 27, 1794, a Senegalese lion (accompanied by his playmate, a young dog), a quagga (an extinct zebra subspecies), and a northern hartebeest (a kind of antelope, now also extinct) arrived at the Jardin des Plantes. The hartebeest died of injuries sustained during the transfer. A month later, the creation of the Ménagerie du Jardin des Plantes was made official by a decree issued by the Convention, and dated June 16, 1794. At that time, the Ménagerie already housed sixty-five mammals and twenty-five birds. The same year, the former animal tamers (notably Félix Cassal, Dominique Marchini, and Bernard Louzardi) were appointed as keepers.

Close of the Century at the Jardin des Plantes

The plot of land where the Ménagerie is now located was acquired in 1795. A year before, the museum had designated Jacques Molinos as its architect, and he set about building pavilions and "primitive" huts in the park. Between 1796 and 1798 the armies of France confiscated numerous specimens as trophies during their victorious campaigns. The Asian elephants Hanz and Parkie (known as "Marguerite") were seized from the animal park of the Prince of Orange, William V, and in Bern (Switzerland) the army helped itself to some of the city's iconic bears. Some animals were offered as gifts by different nations, while others were brought back from distant countries. For example, ostriches, camels, gazelles, and two lions (Marc and Constantine) were collected in North Africa, under the supervision of Cassal, who had become the Ménagerie's concierge. By this time, the Ménagerie was home to some hundred animals of thirty-seven different species. It also included aviaries, a pheasantry, a dairy, beehives, ruminant enclosures, and ponds. In *Voyage au Jardin des Plantes,* Louis-François Jauffret described a Barbary macaque who obeyed orders given to him, although the same monkey had injured Cassal, by attacking his nose. Jauffret also related how a mandrill kissed him through the bars when he asked him to, and we learned that a lion originating from Senegal and arrived in the Royal Ménagerie in 1788, had recently died, as well as the "grand mandrill." Finally, the author compared the Ménagerie to a prison, where the animals did not have enough space, lacked physical activity, and received poor-quality food.[6] The artist and naturalist Jean-Baptiste Audebert came to the park to observe various primates displayed there. Between 1797 and 1800 he published his *Histoire Naturelle des Singes et des Makis,* a milestone in the history of primatology. The work was illustrated with lavish color plates engraved by the author, mainly showing animals in a natural context. His meticulously precise descriptions include an orangutan and a chimpanzee. At the time, apes posed a problem for the Ménagerie managers, as some of their habits were deemed morally reprehensible: *"One should avoid showing them to ladies or children, especially at certain times when the primate seems to take pleasure in displaying his lust."*[7] At the beginning of the nineteenth century, under the auspices of Napoleon I[8] and directed by Étienne Geoffroy Saint-Hilaire, construction of several buildings was initiated in a bid to restore part of the pomp of the Ancien Régime. The new buildings were constructed following an overall plan proposed by Molinos, including a rotunda (shaped like the Légion d'honneur, the French National Order created by Napoleon in the same year), a palace built for beasts of prey, raptor aviaries, a pheasantry, and bear pits. The park areas were enclosed with fences and planted with trees or bushes. Paths extended through the park and picturesque little cottages and rustic cabins were erected. The only primates present at this stage were a few young baboons. The Ménagerie was open to the public, but the visitors showed little respect for the animals on display. Visitors were prone to let their dogs loose in enclosures or to throw stones at the animals to make them move. Interest was mainly directed toward the big cats.

An Ornamental Garden in the Center of Paris

At that time (which was a period of profound transformation affecting society, mentalities, and ideas), the Muséum of Natural History reigned over the academic world as the undisputed leader, while zoology dethroned botany, the foremost discipline of the Old Regime. By 1828 there were four hundred three animals living at the park, exhibited in accord with biological classification. Among these was the famous giraffe (only the third to make it to Europe) presented to King Charles X by the Viceroy of Egypt, Mehmet Ali. The giraffe attracted countless visitors, all eager to view this rarity. She arrived in 1827 and would survive eighteen years, eventually dying on January 12, 1845. The first orangutan to reside in the park arrived on May 15, 1836. Jack was nine months old and well known for his remarkable ability to imitate human behavior and his custom of eating his meals at his keeper's table. Jack survived less than eight months in captivity. The first chimpanzee exhibited in the Ménagerie arrived on October 18, 1839.[9] He, too, failed to thrive in the park and succumbed only seven months later. Around this time, a new ape house was built by the architect Charles Rohault de Fleury. The press criticized this "palace of the apes" as a rash expense and a foolish luxury. The palatial ape enclosure contrasted starkly with the more rural feel of the rest of the park, where the public could ramble and walk from one object of wonder to the next, discovering rare and curious species. "The Jardin des Plantes was the first example of the union of a picturesque garden with a large menagerie of exotic animals.

This was true even though the wild beasts and the monkeys remained confined to buildings that were not integrated with the landscape, because they were thought to represent another, more scientific and systematic view of nature."[10] An extremely sweet and kind female chimpanzee, named Jacqueline, would benefit from the new monkey house. She is represented by a delicate watercolor on vellum painted by J. C. Werner in 1837, while the orangutan Jack is portrayed in a painting dating from 1836. An engraving also portrays the two apes: Jack is holding a bottle, Jacqueline clasps a carrot in her left hand and a rope in the other. In addition, Jean-Pierre Dantan, known as Dantan le Jeune, crafted a bronze bust of Jack, in the style of Ancient Egyptian statues, in 1836. Both Jacqueline and Jack died of tuberculosis, exacerbated by the inappropriate climate, diet, and habits.

The Rise of Science in Zoos

Under Frédéric Cuvier, officially appointed on December 21, 1803, the Ménagerie became a major research center. In the nineteenth century, it was the only science center that was truly involved in studies of animal behavior and psychology when classification of life forms based on taxonomic research[11] was at its peak. Naturalists then concentrated predominantly on describing specimens in terms of anatomy, physiology, morphometry, and so on. Their main goals were to compare species and build up permanent natural history collections, allowing knowledge to be preserved along with the specimens. Identification of species was an ever-present issue. Descriptions available in the scientific literature were not always adequate for recognizing or naming species with any precision. Accordingly, it was often difficult to identify the species for certain residents of the Ménagerie. Furthermore, practically nothing was known about the natural diet, illnesses, or daily requirements of the various species on display. After Cuvier's death, Isidore Geoffroy Saint-Hilaire continued research in a laboratory built in 1846 in the center of the park. Much like his predecessor, he considered the great apes to be particularly worthy subjects of scientific study. During the Siege of Paris in 1871, several animals were slaughtered and eaten, notably the antelopes, the camels, and zebras. The highly popular elephants Castor and Pollux (visitors would line up to be carried around the Ménagerie on their backs) also ended up on the dinner plate. In contrast, the primates were spared, as they were deemed too close to humans.

Splendor and Decline of the Ménagerie

A crocodile room, a reptile gallery, a new pheasantry, an otter pool, and a large aviary were constructed between 1870 and 1888. The aviary, designed by Alphonse Milne-Edwards, was built for the 1889 universal exhibition and remains to this day one of the most impressive buildings of its kind still standing in Europe. A gorilla arrived in the Ménagerie during the year 1884. The animals were displayed in an enchanting and varied setting, providing the inhabitants of Paris with a place to wander and enjoy "a prophylactic dose of nature"[12] within the urban environment of the city. The Ménagerie enjoyed a period of splendor and evoked great interest from the scientific world. Zoological parks had long been symbols of power for influential people. Rare animals were valuable possessions displayed to great dramatic effect. During the colonial expansion in the nineteenth century, such exhibits served to showcase how colonized peoples across the world were dominated by the leading European nations. At the end of the century there were 1,300 animal residents at the Ménagerie. But their living conditions began to deteriorate and competition from many other zoos became a factor. Although wood stoves were installed in the enclosures of exotic species, primates were continually threatened by the cold. Being familiar only with tropical climates, they were simply unaware of the need to protect themselves from the cold, and could not do so in any case. Examination of zoo records dating from 1837 to 1965 revealed that on average primates generally survived for only about 18 months in captivity. Some lists found in the "Mammals and Birds" section of the Muséum show that different apes (thirty-five individuals) that arrived from French West Africa, Gabon, Borneo, and Sumatra (1897–1961)[13] had a life span ranging from three months to thirteen years for orangutans, while chimpanzees survived between eight days and twenty-eight years.

The Ménagerie in the Twentieth Century

A photograph taken in 1903 shows the young chimpanzee Édouard snuggling in the arms of his keeper, André Nicolas. In those days, primates were given local human first names. But surnames were never used as they were reserved for people. Nowadays, exotic names are far more fashionable for primates kept in captivity. Around 1910, there were 1,700 animals, including 407 mammals, at the Jardin des Plantes. They included some fifteen mandrills, but no mention is made of anthropoid apes. At that time, the Ménagerie was plagued by a series of major problems. The buildings were thoroughly antiquated, zoo employees were not specially trained, there were far too many animals for the space available, and the animals' diets were inappropriate. The captive populations were decimated by multiple insults such as cold, humidity, and inadequate hygiene. After serving as a leading example for so long, the Ménagerie was surpassed by other zoos. For example, at this time Hamburg Zoo was inaugurated, including a luxurious building specifically built for primates, the largest of its kind. Improved knowledge of natural habits permitted more informed planning of enclosures, which were built to meet the actual needs of individual species, rather than merely to enhance the enjoyment of zoo visitors. Examples of this new kind of enclosure were the large exterior space available for the lions at Berlin Zoo, the vast tree-filled aviary in Rotterdam, and the island and pool for seals in London. In 1907, the German Carl Hagenbeck ("the most prominent collector and merchant of wild animals in history"[14]) inaugurated a truly revolutionary zoo in Stellingen, near Hamburg. Visitors to the zoo could view the animals as if in their natural habitat in areas without cage bars. Thereafter, zoological parks were seen in a radically different way. In particular, zoos began to make a concerted effort to recreate the natural habitats of the species in their care. This major shift in thinking eventually led to the construction of an "African savannah" through which visitors were able to travel aboard a small train, in the San Diego Zoo Safari Park. This park, linked to the San Diego Zoo, was built at the close of the twentieth century at Escondido in San Pasqual Valley, California. It extends over an area of 1,800 acres, half of which consists of protected natural habitat areas.

Displaying the Natural World

After the First World War so few animals were left at the Ménagerie that the Muséum was obliged to organize collecting expeditions in the French colonies. In 1930 the famous animal sculptor François Pompon, who had been Auguste Rodin's assistant, portrayed one of the menagerie's orangutans in the style of ancient Egyptian statues. According to a report by Bourdelle, in 1931 there were several anthropoid apes among the 317 mammals, which represented 110 different species. There were four gorillas from Cameroon and the Ivory Coast, as well as eight chimpanzees of two different subspecies, also from Ivory Coast. Orangutans were also included: Paul Rode and Edouard Bourdelle wrote an article about an orangutan in the Ménagerie.[15] A new ape house, designed in the Art Deco style by François-Benjamin Chaussemiche, was constructed in 1934. But the opening of the Zoo of Vincennes (now Zoological Park of Paris[16]), some three times larger, in the same year overshadowed developments at the older Ménagerie of the Jardin des Plantes. Nevertheless, a new big-cat house was built there in 1937, replacing an older construction built in 1921 and symbolizing the institution's renaissance. In June 1939, ten years after the species had first been officially described (Schwartz, 1929[17]) a young bonobo (Kitoko) arrived in Paris. But it was sent to the zoo in Vincennes and not to the Ménagerie. Sadly, the bonobo died in August the following year. Still, in 1960, visitors to the Ménagerie were able to see all the other ape species (chimpanzees, gorillas, orangutans, and gibbons), in the Fifth Arrondissement right in the middle of Paris. Numerous monkey species were also on display: mangabeys, guenons, baboons. After a long period of neglect during the nineteenth century, hygiene was now a priority for many zoo managers. Because of their susceptibility to infectious diseases such as tuberculosis or influenza, primates were considered particularly vulnerable. Their cages were isolated from the public and subjected to strict hygiene regulations. It was, for example, forbidden to give primates objects to play with. Moreover their diets were carefully designed to include fruits, vegetables, milk, eggs, and meat. The bear pits were also restored in 1960.[18] Clad with large grey pavestones and vaulted niches, as well as metal bars, the pits were embellished by adding a pool, trees, and inside dens.[19]

Breeding Plans

Various zoos began to found associations: the Association of Zoos and Aquariums (AZA) in the USA in 1924 and the European Association of Zoos and Aquaria (EAZA) in the 1990s. Zoos began to justify their existence by highlighting their activities promoting species conservation, their programs of scientific research, and their educational mission. The Washington Convention, signed on March 3, 1973, and ratified by the Ménagerie in 1978, represented a major advance in the realm of animal protection. At last the world was provided with a set of clear guidelines and regulations concerning trade in endangered animal and plant species. It was henceforth illegal to capture anthropoid apes in their natural habitats. The convention, established under the aegis of the United Nations, promotes careful management of the genetic resource represented by captive animals. The resulting breeding plans had dramatic consequences for the great apes, who are "eminently social" beings.[20] Because priority was given to genetic diversity and reproduction, individual apes were included in breeding programs and transferred from one institution to another. But these transfers were carried out without considering the social implications for the apes themselves. Relationships between family members, or with other conspecifics, or with their keepers, were simply shattered. Breaking these bonds was particularly tragic, as these captive apes already had access only to very few social partners compared with the natural situation in the wild. Yet the disruptive effects of such separations may vary from case to case. The Dutch keeper Leo Hulsker (Apenheul Primate Park, the Netherlands) explains that transfers are not so problematic for orangutans as changes in partners or location are more likely in the wild because of their social system.[21] Additionally, female chimpanzees and bonobos disperse from their natal group once they are sexually mature. Gorillas of both sexes do the same. Changes in partners thus also happen under natural conditions. Nevertheless, once adolescent or adult apes are integrated into a group they usually establish long-term bonds. Apes maintain contact with family members and significant "friends"[22] for most of their lifetimes. Gorillas frequently form long-term pairs and these relationships occasionally last for life. Koko, a "talking" gorilla,[23] behaved with great determination while choosing a partner from photographs, stubbornly refusing some males and clearly preferring others. In zoos, partnerships are initially imposed and then frequently disrupted. Separations between primates and their keepers are often traumatic because of the typically tight bonds that become established between them.

Zoos, Keepers, and Great Apes

When Gérard Dousseau started as a keeper at the monkey house of the Ménagerie (where he is now head keeper), on September 14, 1974, he was in charge of two orangutans (Nénette and Toto, both about five years old), three chimpanzees (the male Cléo, born in 1967, and two females, Pamela, born in 1963, and the fifteen-year-old La Vieille), and three gorillas (the males Horatio and Néro, and the female Valie, all kept in separate cages). Direct contact sometimes took place between Gérard and Valie, who was bottle-fed by her keepers and hence accustomed to human contact.

At the time, anthropoid apes were exhibited in isolation, one after the other in a row of cages: a solitary gorilla in an empty cage, a chimpanzee deprived of the company of conspecifics, an orangutan squatting alone on a tiled floor. These were primates without families or things to do, victims of the hygiene-driven thinking that dominated at the time. Dousseau recounts how Horatio was locked up, alone and with a chain cemented into the ceiling as his only accessory. So Dousseau gave Horatio a tire to keep him occupied. The gorilla kept it for several days, despite his species' general aversion to any foreign object. This shows precisely how starved of distractions Horatio was. Yet the keeper was reprimanded, because it was against the rules to give the primates any form of toy, as they were "not there to play"[24] but to serve as exhibits. It should be noted, however, that the notion of the zoo as a *collection* still prevailed at this Parisian institution in the mid-twentieth century. The Ménagerie housed the living specimens, whereas the Muséum served as a depository for the natural history collections containing dead specimens. As a consequence, the social life of primates was not high on the list of priorities. But gorillas live in small, relatively stable family units, including several females and their young protected by an experienced male. The silverback is prepared to defend the family even at the cost of his own life. Being isolated in an empty cage must have been an unbearable ordeal for Horatio, and for all the many gorillas that have shared the same fate.

Gérard Dousseau was also responsible for looking after the chimpanzees, and was particularly impressed by the extremely intelligent female Pamela. This chimp had already spent time with human beings before arriving at the Ménagerie in 1972, and had picked up some distinctly human habits, such as smoking. She could actually light a cigarette by placing it end-to-end with an already burning cigarette. She was also an avid beer drinker and would dress up if given human clothes. She could wear spectacles and would clean the lenses when they became dirty. When the keepers gave her a mirror, she would instantly recognize herself and use it to check her looks. The keepers also used Pamela as the subject of a "radiophonic experiment," giving her access to a transistor radio on the other side of the cage bars. Pamela was quick to learn how to switch the radio on, and could even turn the dials to find a suitable frequency. She showed clear sensitivity to the subtleties of sound, both in music and in speech. Pamela died in January 1985.

Gorillas and Chimpanzees

The last gorilla to be housed at the Ménagerie died in 1982. The four chimpanzees that were still living there, La Vieille (second of the name), Igor and Charlotte (who arrived in 1987), as well as their daughter Domi, were sent to the Beauval Zoo in 1992. The Dousseau family planned a visit to the chimps five years after their transfer, arriving at the zoo on a rainy afternoon. They approached the chimpanzee enclosure where fifteen individuals were sleeping (it being siesta time). As soon as they arrived, a trio of chimps peeled away from the group and came to look through the window, directly at the two animal keepers. They were the chimpanzees from the Ménagerie. A further anecdote serves to underline the capacity of great apes to recognize people who have been especially close to them in the past. During his annual holiday, Gérard decided to shave his beard. He had worn a beard since his arrival at the Ménagerie as an eighteen-year-old. When he returned to the ape house for the first time after his vacation, Nénette (who had always known Gérard with a beard) was clearly puzzled. It was only when she heard Gérard's voice that she was willing to descend from the upper level. She then approached him slowly, watching him carefully, and held out her hand. She touched him cautiously and then sniffed her hand. She had now verified that this man was indeed her favorite keeper! Primates hence "superimpose" various items of information acquired through their senses of vision, smell, touch, and hearing, and use these elements in order to identify people they are faced with. It is clear that primates can recognize people they have close bonds with, even after very long periods of time, or after their physical appearance has changed. Leo Hulsker, the keeper at Apenheul already mentioned, also has an example in this vein, demonstrating apes' surprising ability to recognize familiar faces. Leo tells the story of how the male orangutan Karl immediately recognized his old keeper when he came to visit. Karl was at the back of the cage but rushed to the bay window, where Leo was waiting. The orangutan had been fond of his keeper when they were together at another zoo, thirty-seven years before Leo's visit! Karl was only eight or nine at the time, and had not seen his keeper again since then. Great apes have an extraordinary ability to discriminate even the subtlest differences in facial traits and physical characteristics. This keen capacity functions in tandem with a powerful long-term memory, something that is of particular importance to orangutans living in the wild. Individuals living in the same region encounter each other only rarely, and these encounters occur at long intervals. It is crucial for wild orangutans to be able to distinguish different individuals very rapidly in order to adopt the right response: caution, empathy, or escape. This impressive memory capacity is also a valuable asset for dealing with the bewildering diversity of plant life in the forest. Orangutans need to know the many plants and their fruits, their characteristics and properties, and the location and timing of fruiting. Some authors have suggested that orangutans might need to be familiar with about a thousand plant species. Others place this estimate much higher, closer to four thousand species. In any case, we often underestimate the extent of primate knowledge, as the concept of "knowledge" is strictly reserved for humans.

Living at the Zoo

Starting in the 1980s, a program of improvements was established to rehabilitate the Ménagerie. Several buildings were renovated: the rotunda, the raptor aviaries, the large aviary, the big-cat house, and the reptile house. The park regained much of its charm, becoming a beautifully kept zoological garden while retaining the aura of its bygone days. It then harbored more than a thousand animals maintained by some sixty staff members: director, curators, veterinarians, keepers, technicians, and supervisors. Zoological parks at the time pursued their goals of displaying natural objects in a cultural "dispositive" (in the Foucauldian sense[25]). Assuming a wide range of variation according to the spirit of the times, this shared desire to exhibit nature was driven by humankind's basic will to affirm supremacy over all other living things. A consequence of this attitude is a notably casual attitude toward locking animals away.

All the same, zoo directors and keepers did begin to show a gradually increasing concern for the welfare of animals in their care. Seen from this point of view, the notion of environmental enrichment (born in the USA) represents a decisive breakthrough. People were finally giving thought to providing captive animals with interesting activities, and hence enabling them to escape the boredom that so often plagues zoo animals. The goal of enrichment is to improve multiple aspects of the animals' environment and living conditions, at different levels: the cage and its elements, the animals' social life, cognition, nutrition and foraging, technological environment. Such improvements are especially beneficial for primates.[26] At the Ménagerie, the managers increased the number of cages available for the orangutans. An outside enclosure was built, so that the orangutans could spend some time in the fresh air, as well as benefiting both from more space and from access to sunshine during warmer weather. The orangutans were also given access to many accessories to pass the time. These included objects to encourage aquatic games (inflatable pool, bucket, watering can), jute bags, toys, clothes, cardboard tubes, branches, food mazes, and various other things. Artificial termite nests were also provided. Much like free-living chimpanzees and orangutans, they were then able to use stems as tools, not to extract termites or ants as in the wild, but to collect molasses or honey mixed with dried fruits (*enrichment of manipulation and cognition*). Structures in their enclosure, such as the wooden gantries, ropes, or tangles of fire hoses, were also modified regularly (*structural enrichment*). Some zoos (including Fresno Chaffee Zoo, San Diego Zoo, Apenheul Primate Park) also took the step of allowing different species to live together in the same enclosure. For example, orangutans shared enclosures with siamangs or langurs (*social enrichment*). The keepers also provided apes with manual activities such as painting sessions (*interactive enrichment*). Moreover, they would sometimes hide the primates' food or make access to food more difficult, in order to increase the time spent "foraging" (*nutritional enrichment*). In tropical forests, orangutans in fact spend an average of seven hours a day foraging and feeding. They will eat about eight pounds of fruits[27] and leaves, along with bark, roots, flowers, sap, mushrooms, honey, eggs, young birds, caterpillars, and a variety of insects. Occasionally, they also hunt small-bodied primates.

Orangutans at the Ménagerie

Since the Second World War, twenty-one orangutans were reportedly housed in succession at the Ménagerie of the Jardin des Plantes. A male from Sumatra who remains nameless in the *Studbook* was brought to Paris in 1938 and resided there for the duration of the war. He died in 1947. In May 1949, it was the turn of the male Peet and the female Greetje to be residents of the Ménagerie. They were captured on Sumatra, transported to Rotterdam, and then transferred to Paris. Their age was unknown. Peet died in 1951, followed by Greetje in 1953. Tubor arrived in Paris in November 1954. He lived alongside Nénette, who arrived at the zoo at the same time as him. Today, a female "Nénette," the matriarch of the Parisian orangutans, is well known to all the Ménagerie's visitors. At about 44 years of age, she is the unchallenged queen of the small tribe of females living at the Museum. Currently, there are only three living there: Nénette, Théodora, and her daughter Tamu, each representing a different generation. We will discuss them later in the text. A male, Joey, 18 years old, came from the Sóstó Zoo in Hungary and joined them in November 2013. Unfortunately, affected by severe digestive problems, he died on July 23, 2014. Few people are aware that today's Nénette is actually the second to carry the name, and is in fact recorded as "Nénette II" in the *Studbook*. The "first Nénette" had resided in Paris from November 1954 until her death in December 1965. After the death of the male Tubor in June 1959, Henri became that Nénette's new companion. He lived up until September 1970. Another female, Pierrette, lived at the Ménagerie from August 1960 to September 1971. All of these orangutans came from Sumatra, but some more recent orangutan arrivals originated in Borneo. Toto (captured in 1972[28]) and Nénette II (caught in 1969) began their residence in the French capital in June 1972. Seven years later, Nénette gave birth to her infant Doudou, who was the first orangutan born at the Ménagerie, on June 22, 1979.[29] As was the case with all four of her sons, she proudly presented the newborn baby to her keepers, Gérard and Danielle Dousseau. She showed them the baby through the window as soon as they arrived at the ape house, further proof (as if it were needed) of the profound attachment to her keepers.

Of Deaths, Births, and Transfers

The female chimpanzee Chloé was born in the Vincennes Zoo on December 13, 1980. She was hand-reared and was brought to the Ménagerie on May 5, 1981. She was reared with the young orangutan Doudou, who joined her when he was two-and-a-half years old. Chloe and Doudou were raised together and were involved in a three-year research program led by two young scientists. Bernadette Bresard carried out work on laterality and differentiation/coordination between the two cerebral hemispheres, through observations and experiments on right and left asymmetry (notably related to the functions of the hands) with the two young apes.[30] The other researcher, Marie-Christine Lacour, focused on issues of nonverbal communication, such as interspecies communication and expression of the emotions based on facial features, and on the question of the recognition and dissymmetry of facial expressions.[31] During this study, the two women noticed that the two apes imitated certain postures and gestures between themselves. Doudou and Chloé were subsequently transferred to Kyoto, Japan. The male, barely an adolescent, died there five years later.

Toto and Nénette had another son, Mawa, born on February 22, 1983. Mawa was not even three years old when he was moved to the Natura Artis Magistra, or Artis Royal Zoo, in Amsterdam. He stayed there until his death at the age of twenty-four in 2007. At that time, young apes were taken from their mothers as early as possible, to speed up subsequent breeding. As a result, it was typical for young apes to be raised by their keepers. Toto died in February 1985. In September of the following year, Papou was brought to Paris. His hindquarters were paralyzed and he died two years later. In May 1987, two young females arrived at the Ménagerie, following confiscation by customs officials at Charles de Gaulle airport, who had no idea which species the apes belonged to. One of the females was estimated to be three years old and the other as one-and-a-half years of age. The younger one, named Ralfina, is of particular importance as far as this book is concerned. She was the orangutan who would eventually give birth to Wattana in 1995 at Antwerp Zoo. Ralfina had been transferred there one-and-a-half years after her arrival in Paris. The other orangutan from Charles de Gaulle airport, Ralfone, was repatriated and released in Sepilok (Indonesia) in June 1990. Nénette's third companion, Solok, arrived in Paris in December 1987, a short time after the death of Papou in June of the same year. Solok is a magnificent male, weighing in at 240 pounds and endowed with an impressive facial disk. Solok and Nénette went on to have two sons: Tubo was born on May 15, 1994, and his brother Dayu just over five years later, on December 6, 1999. Both were raised by Nénette. Before this second birth, however, another event threw life at the Ménagerie into upheaval: the imminent arrival of two young orangutans, transferred from the nursery at Wilhelma Zoological-Botanical Garden, in Stuttgart.

Figure 1.1 Wattana on her arrival at the Ménagerie du Jardin des plantes, courtesy of Frank Simonnet, French National Museum of Natural History, Paris, 1998.

Wattana and Vandu Arrive in Paris

The ape house of the Ménagerie was originally built in 1934 and later refurbished. It was here, more than 60 years later, that Wattana and her half-brother Vandu were welcomed on May 5, 1998. The young female was not a complete stranger to the keepers. As already mentioned, ten years earlier they had taken care of Wattana's mother, Ralfina. Margot Federer, the German keeper who had cared for the two orangutans, had also been sent to Paris to facilitate the transition. After having spent a little more than two-and-a-half years in Germany, with apes and human beings she had grown deeply attached to, Wattana was forced to adapt to new surroundings: new faces, new smells, different habits, and novel enclosures. This was an environment very different from that of the nursery at Wilhelma Zoological-Botanical Garden, in Stuttgart, where she had been reared. In Germany, the two orangutans lived in a very human little universe. They had access to toys, cradles, and playpens. Additionally, Wattana and Vandu

Figure 1.2 Wattana sleeping in the arms of her brother, Vandu, at the Ménagerie, courtesy of Frank Simonnet, French National Museum of Natural History, Paris, 2000.

were used to being cared for by female staff. In Paris, by contrast, the strangers who surrounded them were predominantly male. To make things even more challenging, these strangers spoke French, a language unfamiliar to the orangutans.[32] Furthermore, the adjacent cage was occupied by three other orangutans, including two adults. This was an unfamiliar situation as previously they had encountered only very young apes. Their life was hence radically changed. Despite the great extent of all these changes, Margot Federer stayed for only a few days. As explained above, bonds are broken from an early age in zoos, whereas in the wild mother and offspring may stay together for up to nine years, and may meet up again later during their wanderings in the forest.

Connected by Trust

Once Margot had left, Vandu was no longer content with simply being a playmate for Wattana. He became her protector, despite being only three-and-a-half years old. The two orangutans slept curled up together each

Figure 1.3 Wattana drinking her bottle of milk, courtesy of Frank Simonnet, Paris, 1998.

and every night. They would awaken at the crack of dawn and await the distribution of food. The young female still drank three bottles of milk a day, while Vandu would drink his milk from a glass. Their diet was supplemented with fruit and vegetables, occasionally with the addition of yogurt. The keepers then started to add jars of baby food and pet food pellets, entering the enclosure every day to feed, cuddle, and play with the two small primates. In this urban zoo founded at the end of the eighteenth century, the historical buildings were all protected, so the available space was tightly limited in comparison to more modern zoos. Possibilities for modifying existing structures or replicating natural environments were also severely constrained. The keepers hence had to compensate for the space restrictions and be resourceful to ensure a certain standard of well-being for the animals on display. Wattana progressively became accustomed to her keepers. She was particularly fond of two of them: Valérie Martinez and Franck Simonnet. Franck was very close to the two orangutans, spending a lot of time with them, and was often seen walking around with the two little ones in his arms. Valérie was both a friend[33] and a surrogate mother for Wattana. Tight bonds were formed and became evident through both looks and gestures. For example, the orangutans would slip straws through door hinges or railings when the keepers were passing by. It was then vitally important for the keepers to return the straws, as, for the apes, it would surely have been particularly impolite not to reciprocate. This game could carry on for a while and consist of several such exchanges. Unfortunately, Franck left the Ménagerie to take up a job elsewhere. Wattana experienced yet another separation: in addition to apes' transfers between zoological parks, keepers are often likely to move as well.

Tubo, Vandu, and Wattana

Towards the end of 1999, Nénette and Solok's son Tubo was gradually introduced to Vandu and Wattana's cage. During such changes, keepers must draw upon their wealth of experience and their often subtle but in-depth knowledge of the animals concerned. The keepers' remarkably honed ability to make judgments, particularly concerning how social groups are organized, is invaluable to zoo curators and managers. After this period of change, the young male remained in visual contact with his family members, although they lived in separate cages. Once united, Tubo, Wattana, and Vandu engaged in intense play sessions, spending most of the day pursuing or wrestling with each other, locked in fraternal embraces. These activities allowed the orangutans to profit from physical exercise to increase their muscle power. Furthermore, through one-on-one interactions they learned to develop healthy social relationships. It also happened on occasion that they would grasp their keepers with arms or legs to force them to stay with them. These early signs served as a warning to the keepers and prompted them to limit time spent inside the enclosures.

Figure 1.4 (From left to right) Vandu, Wattana, and Tubo, courtesy of Frank Simonnet, French National Museum of Natural History, Paris, 2000.

Then, in order to promote genetic diversity and avoid confrontations between the two males, the zoo managers decided that it was time to move Vandu to another zoo. To prepare Vandu for this separation, he was isolated in a cage facing that of his former companions. His isolation began in April 2001 and was unexpectedly extended because of administrative delays. Wattana hence continued to see Vandu from afar. After spending every moment of the day with her half-brother, for over three years, Wattana was now alone. The keepers recounted how the young female clearly expressed her sadness. She spent much of her day crying and whining. They also observed how she would clutch a small bundle of straw to her body every morning.[34] Vandu left the Ménagerie for the last time on August 26, and was moved to the Sóstó Zoo in Sostofurdo, Hungary.

2 Growing Up among Humans

Quite often I had to remind myself that this little chimpanzee girl was not a human being. But after a while I realized that this distinction had become meaningless to me.

—ROGER FOUTS AND STEPHEN TUKEL MILLS, *Next of Kin*, 65

Before arriving at the Ménagerie of the Jardin des Plantes at the age of two-and-a-half, Wattana had already experienced life at two other zoos. She was born in Antwerp Zoo, in Belgium, on November 17, 1995. Because her mother, Ralfina, was unwilling to look after her, she was placed in the care of her keepers. Wattana was subsequently transferred to Stuttgart Zoo. At the time she was barely three-and-a-half months old. Ralfina's attitude towards her daughter might be explained by the fact that she was only ten years old[1] when she gave birth. Nevertheless, in their natural habitat, some orangutan females are not much older when they give birth for the first time. Several sources do, however, state that adulthood corresponds to an age of 14 or 15. In fact, there is no *standard* age for becoming a mother, beyond the numbers generated by scientists (which vary greatly between sources anyway). Science

Figure 2.1 The mother of Wattana, Ralfina, who never learned to be a mother.

aims to establish generalizations, yet no individual orangutan is in itself a generalization. The duration of childhood, age at sexual maturity or first pregnancy, the interval between births, and life expectancy all vary between individuals, between subspecies, between different species, and from one distribution range to another.[2] For example, it seems that the time that elapses between two births for a given female is roughly three to four years in *Pongo pygmaeus morio*, yet seven, eight, or even nine years in the other subspecies. Furthermore, individuals can differ markedly from one another. Among rehabilitated orangutans, groups of friends are formed in a manner reflecting the skills of each group member, and as a function of how they complement each other: one member may be skilled at constructing nests, while others may be expert foragers, or have a knack for detecting dangers. They thus constitute a highly effective team with an improved likelihood of returning successfully to their natural habitat if they stay together.[3] As regards mothering, some females are ready at the age of ten, while others need much more time. In any case, Ralfina later failed to take care of any of her three subsequent offspring: Binti (a daughter born five years after Wattana, when Ralfina was adult), Sayang (a son who was born in 2003 but died two weeks later), and Dayang (another daughter, born in 2005).[4] Ralfina did seem to be prepared to take care of her last daughter, but was unable to breastfeed her. In order to prevent the baby from starving, the keepers successfully managed to contrive her adoption by another female,[5] Sandakan, who had given birth to a baby, Samboja, six months earlier. Sandakan thus took care of two infants at the same time. This case of adoption is not uncommon among great apes. While some females very clearly turn their backs on motherhood, others are strongly attracted to babies. Some go as far as taking care of three infants simultaneously.

On the Difficulties of Motherhood

If age is not a factor determining the aptitude for mothering, how can one explain why some female apes are so lacking in a "maternal instinct" when living in captivity? A priori, it would seem that such females should be able to link together the sequence of actions necessary for looking after their newborn babies, as is the case with female cats, for instance. Even if they have no experience at all, cats *instinctively* know what needs to be done after the birth of their first litter. In fact, great apes are envisaged as the natural counterpart to humankind, as authentic *creatures of nature*, entirely governed by *instinct* and *subject to biology*. Accordingly, they should know straightaway how to respond under all circumstances. Yet repeated cases of ape mothers in zoos that show no interest in their babies, or do not know how to look after them,[6] clearly show that the process leading to "becoming a mother" is complex. Moreover, this suggests that the notion of "maternal instinct," as something both universal and automatic, and as the only foundation for this kind of capability, needs reevaluation for great apes. In contrast with numerous mammals that know right from the start how to mother their offspring, great apes must first *learn* how to do this.

Existences

The interval between births and the duration of coexistence with the mother is between three and six years for gorillas, chimpanzees, and bonobos. It is between three and seven years for Bornean orangutans and between eight and nine years for orangutans of Sumatra. Great apes hence benefit from a long infancy that lasts several years. Indeed, the transmission of the whole palette of skills needed to live their lives in the forest environment, and together with other members of their species, requires a long period of training, which lasts roughly seven years in orangutans. Furthermore, their life expectancy, a little more than forty years in their natural habitat,[7] allows older individuals to pass on their experience to younger ones, even if only by providing an example. Such longevity combined with extended infancy characterizes the five kinds of anthropoid primates: bonobos, chimpanzees, gorillas, orangutans, humans. During the first five years of their infancy, orangutans are particularly inquisitive and show remarkable learning capacities, which persist when they become *juveniles* (at between five and eight years of age). According to scientists, adolescence continues until the age of approximately thirteen in males and fifteen in females, a period during which individuals acquire their independence and enter the subadult stage. It is only at the age of fifteen for females, and eighteen for males, that orangutans are considered to be adults. It is predominantly mothers, serving as fundamental relay stations during the training process and as impressive conveyors of culture, who transfer to their young the assemblage of knowledge and practical skills needed to truly *become an ape*.

Learning to Be a Mother

Mothering is also transmitted from mother to daughter. To paraphrase Simone de Beauvoir, in *The Second Sex*: "a female great ape is not born a mother, but becomes one." Orangutan females that are "naïve," meaning those that have not yet produced young, are initiated into maternal behavior by the example of their own mothers, as well as by females encountered fortuitously in the Indonesian forest. Later on, they try to imitate their models. Captured at less than two years of age, Ralfina did not have the opportunity "to learn to be a mother." She did not have sufficient time to soak up her mother's influence. Attempts are sometimes made to *show* inexperienced females how to look after an infant by providing cuddly toys or allowing visual contact with primates of other species that have recently given birth. But here zookeeper Leo Hulsker thinks that it is absolutely critical to provide an appropriate model for mothering: to learn maternal actions, an *orangutan* must have an *orangutan* mother as an example. The process of projection and identification is then easier. In any case, this approach is rarely possible, as pregnant great apes are a rarity in zoos. Furthermore, even when they do have the opportunity to learn from an effective model, young female apes are still deprived of all that is transmitted through the close proximity of bodies: a particular maternal style will influence the baby more surely if it is *incorporated*, experienced in the flesh and picked up by all the senses. It is while growing up close to its mother (and even tightly clinging to her) that an infant acquires the knowledge essential for living in its world.

Becoming an Orangutan

Training may be accomplished in a *passive* manner, via imprinting; *actively*, through imitation; or in a *doubly active* fashion (a form of teaching that is exceedingly rare in great apes), with the mother teaching her infant a behavior while it participates actively in learning, focusing its attention and employing his observational sense and its capacity for imitation. The infant then attempts to reproduce the gestures it has been shown. This type of teaching is illustrated by transmission of the skill of breaking open nuts by female chimpanzees in the Taï forest, observed by Christophe and Hedwige Boesch[8]: a mother slowly performs and repeats the necessary steps, while ensuring that the infant is watching attentively. She repositions the nut correctly on the stone anvil when her "pupil" takes its turn to try to break open the nut. An additional example is provided by a mother chimpanzee living in East Africa, who offered a leafless branch to replace the leaf her infant was using to fish for termites. Furthermore, Roger Fouts and his team describe the case of the chimpanzee Washoe, a "talking ape" who taught her adoptive son Loulis sign language using the same demonstration methods as those used by her human teachers. As regards *imprinting*, one of the best examples is of infants being showered several times a day with tiny fragments of fruit and plant matter ingested by their mother, which *passively* introduces them to their future diet.[9] On the other hand, young primates can be extremely active when intensively scrutinizing the actions of their kin,[10] in order to be able to reproduce their gestures through *imitation*. Orangutans use this process to learn how to construct their nests. As they sleep with their mothers sometimes until the age of seven or eight, they are able to witness nest building daily, and carefully study the manner of construction. The young are thus progressively led to try building nests themselves. Thanks to observing their mother and other conspecifics, they also experiment with sophisticated foraging techniques: they learn how to remove prickly appendages from fruits before consuming them, how to extract the pulp from fruits with a hard shell, which insect nests can be explored, as well as how to obtain honey from a beehive high up in the trees using a tool to open it. In this way, they are able to find food items in an environment where they are rare and difficult to access. They learn bit by bit how to judge whether a fruit is ripe, which plants are irritating or toxic, and which plants can be used for healing.[11] They must also know the locations of fruit sources and their fruiting schedules, as fruiting may occur several times a year in some cases, and only once every two to three years in others. Biruté Galdikas declares that orangutans are the "world's finest botanists." Indeed, they truly are "gardeners of the forest," as they contribute to the dispersal of various seeds, sowing them throughout their home ranges. Besides, as infants accompany their mothers on all their travels, little by little they absorb a certain pattern of progression through the trees. In this way, they become specialists of the *three-by-three* method of locomotion, very carefully letting go of only one grip at a time.[12] They also take care to spot which tree trunks might be slippery, as well as judging whether branches are capable of bearing their weight or may break, when they hang beneath them. Orangutans are in fact the most arboreal of the four kinds of great ape, and also live highest up in the canopy. It would therefore be incautious to adopt a style of locomotion that is not secure. Brought up in captivity with young members of other species, Wattana was unable to adopt the type of movement that is typically "orangutan." She would sometimes throw herself from rope to rope, something that never failed to surprise the keepers at the Ménagerie, who were used to seeing a more traditional style of locomotion with their Asian great apes. Finally, turning to an entirely different domain, that of reproduction, great apes must also model their actions on those performed by conspecifics.[13] During the earliest attempts to achieve breeding in zoos, keepers noticed that these primates did not know how to go about it. Ethologists then showed them films taken in the natural habitat, footage that apparently provided the apes with beneficial inspiration.[14]

Cultures and Socialities

Anti-predator strategies, together with production of utilitarian objects and various tools[15] as well as cultural traditions, are also subject to maternal transmission.[16] The mother further initiates her infant into the rules of "orangutan sociality." Orangutans are described as being "semi-social" or having a "dispersed sociality," as inter-individual relationships are spread out over time, and because of the dispersed spatial distribution of individuals.[17] Relationships are formed as a function of encounters, which frequently are chance meetings. In fact, orangutans are not less sociable than other anthropoid apes, but the limited availability of resources simply does not allow several individuals to feed in the same place at the same time. They do, however, have a real need for contact, regardless of whether this takes the form of play involving young ones, communal travel, or sharing time with relatives. All the same, the only stable relationships are those between mother and infant, occasionally in the company of a young adult male, who protects them. A unit of this kind exists alongside others of the same kind or solitary females. Occasionally, a few relatively young females or males will form groups. Additionally, a breeding pair may stay united until conception occurs. Orangutans exhibit a true social intelligence, demonstrating that they are able to interact peacefully, with the exception of encounters between males competing for a territory or a female, or of sexual encounters between males and females: copulations can sometimes be extremely violent. In sum, by teaching her infant to look after itself while coping with the harsh constraints of its physical and social environment, a mother is essentially transmitting nothing less than a user's manual for becoming an ape. This modus vivendi is molded by maternal influences as well as by the habits and traditions of the entire community of orangutans living in a given area.

A Truncated Youth

Torn too early from her own world, Wattana's mother, Ralfina, was unable to profit from this kind of education. In addition, having arrived at the Ménagerie at the Jardin des Plantes in poor health and with a broken arm, requiring a series of plaster casts, she needed frequent and intensive care from the keepers. She was thus deeply *imprinted* by her extensive contacts with humans. Moreover, the foundational (and apparently traumatic) experience of the birth of her first daughter undoubtedly had a major impact, dispatching Ralfina on the trajectory of an existence devoid of motherhood. It is also possible that the young mother was frightened by Wattana. During the onset of the relationship between a new mother and her newborn, the latter is by no means passive, and actually has an important part to play. The newborn will in fact try to kindle interest in the individual that has just brought it into the world, notably by seeking to make eye contact. Clinging to the mother's fur, it will then attempt to crawl towards her teats. Although aimed at grabbing her attention, Wattana's behavior may have seemed menacing to Ralfina, who then became completely disinterested in taking on a maternal role. She is not alone in this. Numerous cases of primates unable to take care of their infants have been recorded in zoos. Yet, if some *female apes and monkeys* do not express the famous *maternal instinct* that is ponderously evoked as soon as the subject of mother-infant relationships is broached, what can one say of *human mothers,* who are above all else conditioned by social and cultural norms, subject to economic constraints and moored by the anchor of symbolism?

Rearing by Hand

In European zoos, young animals have been frequently, perhaps even systematically, schooled through "hand rearing" by humans. But cases of primates adopted by humans have also existed outside zoos. Westerners sent to colonial territories, women and men who founded ape colonies (such as Rosalia Abreu in Havana, Gertrude Davies Lintz in New York, or Mae and Bob Noell in Florida), specialized animal trainers in the entertainment industry (notably those associated with Hollywood), instructors of "talking apes" in the USA between 1960 and 1990, or simply private individuals, have all reared young apes, taking care of them as if they were their own children.[18] In Wattana's case, it was not the first time that the keepers at Antwerp Zoo were obliged to take care of a young ape that had been abandoned by its mother. It was evidently out of the question to "let nature run its course" and allow her to die. For zoo managers, great apes of course represent rare and precious species, to be preserved at all cost. Wattana was hence reared by her keepers and bottle-fed until the age of three-and-a-half months. However, keepers cannot provide a baby ape with the omnipresence and warmth of a mother. In the world of apes, newborns spend all their time with the one that gave birth to them. For the first months, they remain anchored to her stomach. Later, they are carried on her back. As a rule, baby orangutans remain in permanent contact with their mother for more than a year. It is only towards the age of two that they begin to detach themselves and move a few yards away, although they may still drink their mother's milk until the age of seven.

Wattana

Wattana is hence one of a large cohort of apes adopted by humans. Having begun her life as a captive in Antwerp, she is transferred a few months

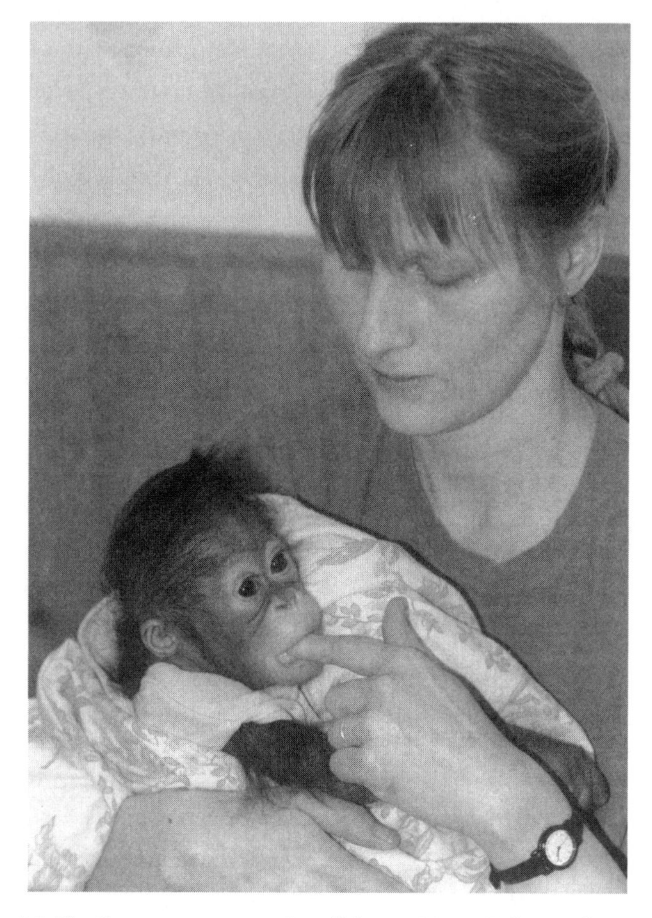

Figure 2.2 The German surrogate mother of Wattana, Margot Federer, with Wattana in her arms, courtesy of Margot Federer, Zoologisch-Botanischer Garten Wilhelma, Stuttgart (Germany), 1995.

later, in March 1996, to the Wilhelma Zoological-Botanical Garden, in Stuttgart. There, in a section exclusively reserved for young apes, she is welcomed by the animal technician Margot Federer. After becoming her substitute mother, Margot quickly realizes that Wattana is extremely intelligent. Wattana immediately grasps what is asked of her, learns rapidly, and shows a particularly keen interest in various human skills and activities. Full of admiration and affection, Margot calls her "my little princess." The keepers play with their little charges in the nursery in the mornings and afternoons. The young apes are able to integrate with great ease into the world of humans: just like human children they wear diapers, sleep in cribs, are bottle-fed, own toys, and are placed in playpens with bars. This education exerted a strong influence both on Wattana's ability to acquire various habits and on the orientation of her ethos, her manner of existence in the world. She was clearly influenced by the ways of life that were passed on to her, and will remain deeply attached to these. Nowadays, keepers try to avoid such practices if possible, in order to avoid having to reintegrate apes that are completely lacking in their own culture into groups whose habits and customs are unfamiliar to them.

Within-Species Interactions

In Stuttgart, Wattana lives alongside young apes of different species and shares in their games. The nursery residents constitute a small community in which chimpanzees mingle with gorillas, orangutans, and bonobos. Using them as models, Wattana adopts some of their species-specific behaviors. If given the chance, apes will imitate each other as much as they imitate humans. The latter also sometimes find themselves in situations in which they are led to assimilate behaviors of the animals they frequent, either through simple imprinting, or deliberately, so as to be able to approach them and understand them better. Dian Fossey used this approach to "slip into the skin" of the gorillas she studied for eighteen years, first in Zaire, then in Rwanda. During her field studies, she crawls along the ground, eats herbaceous *Galium*,[19] imitates the gorillas' vocalizations, and grooms individuals that she knows well. For her part, Jane Goodall tells how she herself became a better mother thanks to one of her favorite chimpanzees, Flo, whom she knew for more than forty years. Isolated on an island in the Gambia, the American psychologist Janis Carter lived, slept, and ate with a dozen chimpanzees for seven years. Her everyday life was hence predominantly shaped by her interactions with the apes. Almost entirely deprived of human contact, she experimented with what I would call "becoming-chimpanzee,"[20] echoing the concept of "becoming-animal" of Deleuze and Guattari.[21] As she herself notes, she really *became one of them*, and it became more and more difficult to *return to being human*. This plas-

Figure 2.3 Margot Federer, playing with Wattana and a baby gorilla of the same age, courtesy of Margot Federer, Zoologisch-Botanischer Garten Wilhelma, Stuttgart (Germany), 1995.

ticity, together with her most precious tool, imitation (so often dismissed with the derogatory term "to ape," as if it were just a stupid and servile form of plagiarism), counts among the most fundamental characteristics of anthropoid primates, including human beings.

Family Histories

Intimately connected with these exchanges of ethos, multiple links enrich the lives of captive great apes, be they with the humans that surround them, other members of their own species, or direct relatives. When an infant is born, a family is formed, and then generations follow one another over the course of time. Accordingly, one can attempt to sketch Wattana's family tree. Tjantikje, the mother of Wattana's father, Tuan, and hence her paternal grandmother, spent all her life at the Cologne Zoological Garden (Germany) with her companion, Maias, Wattana's paternal grandfather. They eluded the shock of transfer from zoo to zoo, but not the shock of exile, as they were both captured in the wild in 1967, she in Sarawak, he

Figure 2.4 The paternal grandfather of Wattana, father of Tuan, photo by Chris Herzfeld, Cologne (Germany), 2004.

in Sabah. Wattana's maternal grandmother was very likely killed when her mother, Ralfina, was captured. If this was indeed the case, the experience must have been traumatic for the baby. In fact, hunters target females that are sheltering in the trees. They then seize any baby once its mother's body has fallen to the ground. The infant thus witnesses the execution of its mother, and is then ripped from her corpse. However, during these bloody scenes, the semi-solitary orangutans suffer less from loss than the other ape species, whose members all live in groups. In these species, capture may entail the deaths of numerous individuals through direct killing or as a consequence of injuries received. The French historian Eric Baratay

Figure 2.5 A magnificent big male orangutan, Tuan, father of Wattana, photo by Chris Herzfeld, Antwerp (Belgium), 1996.

calculated that in the years 1948–49, for every gorilla from the East Congo that arrived alive at a zoo, thirty-two individuals would have been sacrificed, either at the time of capture or later, during captivity in their country of origin or during transport.[22] The German zoologist Bernhard Grzimek estimated that, for each young orangutan captured and delivered alive at its destination, five or six others would have lost their lives. Wattana's maternal grandfather may still be alive and free somewhere in a humid tropical forest of Borneo. It is, however, more likely that he has been killed by a hunter or has since died of natural causes. In the wild, orangutans are now severely threatened and rarely reach the ages of some captive individuals, who may survive for more than fifty years. As it happens, Wattana also has two aunts and an uncle on the paternal side. Indeed, Tuan has two sisters and a brother, all hand-reared: Nonja, Tana,and Muda. Only one of Wattana's relatives lived at Stuttgart Zoo when she was there: her half-brother Vandu, who was also born at Antwerp Zoo, on November 23, 1994. They share the same father, Tuan (who lived at Antwerp Zoo from 1994 to 2007), but have different mothers: he is the son of Astrid, she is the daughter of Ralfina. Vandu was transferred to Stuttgart one month before Wattana was born. They were both raised by humans. Tuan ended up siring nine children, who are hence Wattana's sisters, half-brothers, and half-sisters. The members of this extended family will probably never meet.

On the Way to Paris . . .

As for Wattana, she is perfectly integrated into the small mixed community of humans and apes at Stuttgart Zoo. As a result, she is imprinted in indelible fashion by the human world, and shows an immoderate appetite for the multiple possibilities it offers. But, after spending two years in Germany, she is forced to leave the place where she spent her entire infancy. Her half-brother, Vandu, accompanies her on the second long voyage of her life, this time towards Paris. The orangutans are placed with their blankets in individual transport crates. Then, after a journey of more than seven hours, they arrive in an unknown city where everything seems foreign to them. Nothing now resembles what the two young orangutans had experienced up to that point.

At the Ménagerie

Wattana and Vandu gradually become adjusted to their new way of life. Three years after arriving at the Ménagerie of the Jardin des Plantes, however, the young male is transferred from France to Hungary, leaving his young sister behind. He is six-and-a-half years old, and Wattana a year younger. After his departure, she finds herself confronted with Tubo. The orangutan group at the Ménagerie then consists of five individuals: Wattana; an adult female, Nénette; and three males at different stages of development. Dayu is still a child, Tubo an adolescent, while Solok is an adult. Wise Solok—an impressive sight because of his large size, magnificent red fur, and facial disk—rules over the ape house. He displays great patience towards his second son, Dayu, born in December 1999, and spends a large part of his day playing with him, as he had done with his first son, Tubo. The capacity of the large male to moderate his actions and restrain his strength when interacting with his son is truly remarkable. In the wild, this type of interaction would simply not occur: adult males are solitary and only encounter mother-child pairs by chance. The fact that Solok and Nénette are in constant contact, as they occupy the same cage, has a further consequence. Being much larger and stronger than the female, the male frequently attempts to copulate with her forcefully.[23] The two apes are actually exposed to a totally artificial situation. In the forests of Borneo or Sumatra, although males—generally inexperienced individuals—coerce females into mating, this occurs at far lower frequency. Encounters are only occasional or seasonal. The orangutan is the only ape species that does not live in groups, yet in captivity little concession is made to its social needs or environmental habits. Despite being semi-solitary, orangutans are forced to live constantly with several others. Furthermore, some of them are still presented to the public on an island with a few relatively low-slung wooden structures. Enclosures contain no "real" trees. Although they are the most arboreal great apes, orangutans are not given the possibility to practice their forest-living skills and are instead constrained to live on the ground. In their natural environment, it is rare for them to venture down from the trees.

Within-Species Transmissions and Death of a Giant

Nowadays, it is widely recognized that "hand rearing" engenders several problems, among which is the absence of limits that a parental figure could give: many indeed reckon that Wattana was not sufficiently controlled. Only an orangutan mother, one well prepared for her duties, would have been able to provide the necessary training. In a way, Tubo, now the sole companion of the young female Wattana, became the role model that she had lacked. He had benefited from a more stable infancy than hers, as he had spent his first years at the bosom of Nénette, who would also prove to be a good educator for his younger brother Dayu. Moreover, he had also profited from some degree of bonding with his father. Thanks to Tubo,

Wattana was able to assimilate various "orangutan" codes and habits. She thus applied herself to learning various everyday actions—building a nest, for example. Up to that point, she had simply covered herself with a blanket. At the outset, her rudimentary attempts resulted in quite basic bedding arrangements, some semblance of a nest in which Wattana slept alone. Later on, she managed to construct nests that were just right.

Solok died on February 14, 2004. Previously, the two groups had been separated, but they were now united into one group that occupied the five cages at the end of the ape house, together with the outside enclosure. Thereafter, the keepers would no longer enter the cages. The young orangutans had become too strong, and hence potentially dangerous. Moreover, the presence of Nénette, an adult female, would have rendered encounters between humans and apes even riskier without special precautions. In fact, it is easy to forget that one of the characteristic professional traits of keepers taking care of great apes is that they are faced with a major obstacle: their desire to bond with the primates is constantly overshadowed by the justified fear of being attacked. Because they feel so close to their charges, they would like to increase the frequency of contacts with them and establish a relationship based on trust; yet they know precisely how dangerous such exchanges can be. So they must always remain alert to the danger. Great apes actually possess impressive physical strength.[24] A number of attacks have occurred in zoos, at the homes of private individuals, or in sanctuaries. This explains why, in most zoos, it is strictly forbidden to touch apes or enter their enclosures. The fact that keepers at the Ménagerie of the Jardin des Plantes occasionally have direct contact with their charges is really unusual. In any event, they do so while limiting the risks and in full awareness of the dangers involved.

Christelle and Wattana

When she arrived at the Ménagerie, Wattana had a clear preference for the keeper Valérie Martinez.[25] Yet bonds were also woven with Christelle Hano,[26] who started work at the institution in 1999, one year after the young female's arrival. For example, at the age of five-and-a-half, Wattana

Figure 2.6 Wattana interlacing her keeper, Valérie Martinez, in the cage of the orangutans, courtesy of Frank Simonnet, French National Museum of Natural History, Paris, 1999.

would take her hand to follow her on a tour of her enclosure. Moreover, Christelle had understood the young female's need to think and test, her need to challenge herself, her taste for analysis and thirst for experiments. When Valérie had to leave the ape house and Wattana once again experienced a separation, her bond with Christelle became tighter still, maturing into a selective affinity. Thereafter, Wattana would turn to Christelle when she was ill and wanted to be comforted. That young woman is the only one authorized to enter Wattana's cage in order to treat her. This relationship of trust gives Christelle an aura and authority that is lacking in less experienced keepers, who, as recent arrivals at the ape house, are frequently bul-

Figure 2.7 Wattana and Christelle, her favorite keeper, courtesy of Christelle Hano, French National Museum of Natural History, Paris, 1999.

lied by the female. The apes experience a different chronicle of events with each animal technician. As a result, the keepers serve specific functions in the apes' group of "social partners": playmate, foster father, replacement mother, role model, or troop leader. In addition, the great apes are able to decipher the moods and character traits of their human companions with great subtlety. They are quite capable of evaluating and are perfectly aware of which keepers they can harass or attack, and which they need to respect and with whom it is riskier to disobey. In *The Parrot's Lament*,[27] Eugene Linden narrates the adventures of the orangutan Jonathan, who is an escape specialist, and has hence become, in a way, a "consultant" for enclosure designers. This ape picks locks with a small metal rod that he carefully hides between his gums and lips whenever an experienced keeper is present. In fact, he does not bother with this subterfuge when an unsuspecting volunteer takes care of him, thus showing that he discriminates between different competence levels of the humans he encounters. He thereby confirms his ability to evaluate situations and surmount spontaneous responses. In fact, great apes are excellent observers of humans, shrewd psychologists, and accomplished ethologists. Basing their evaluation on minute signals, invisible to our eyes, they analyze the slightest variation in our voice or mood, odors, body language and posture, facial expression, and the way we move.[28] They thus have access to a large number of keys to comprehension, permitting them to interpret our dispositions.

Adolescent Games?

Following the death of his father Solok in 2004, Tubo shows more confidence. His build becomes impressive, his hair increasingly thicker and longer. He forces Wattana, who is still adolescent and more slightly built than him, to mate. In orangutans and gorillas, sexual size dimorphism is very pronounced. In extreme cases, large male orangutans may weigh up to 200 pounds with a height between four-and-a-half to five feet,[29] while females generally do not reach more than 100 to 110 pounds and three-and-a-half feet. In gorillas, males may reach a height of just under six feet (versus five feet for females) and weigh 440 pounds (versus 220 pounds for their mates).[30] Such intersexual differences are minor in chimpanzees and bonobos. The latter species is the only one to possess a special kind of social organization in which the females have the upper hand, thanks to the alliances they form with each other.

It is only when visitors and keepers have left the premises that Tubo forces the adolescent Wattana to accept copulation in various positions. In the wild, orangutans copulate face-to-face, suspended from branches in the trees. The general public has been led to believe that bonobos are the only primates to use this position, reinforcing the notion that these apes are closest to humans. From the 1990s, this argument was in fact frequently evoked to enhance the status of this species, one so close to chimpanzees that it was deemed necessary to find a few powerful arguments to endow it with a distinct identity. In the scientific community, this perspective had also been highlighted, but in less spectacular fashion, from the 1930s, by the German researchers Eduard Paul Tratz and Heinz Heck. Working on a systematic comparison between bonobos and chimpanzees, at Hellabrunn Zoo (near Munich), they were the first to describe the sexual behavior of pygmy chimpanzees, noting that bonobos copulate *more hominum* (like humans), and not *more canum* (like dogs), as do chimpanzees. But they did not publish their article until the 1950s, and even then the public was not ready to hear of a *Kamasutra* for primates.[31] The world therefore had to wait until well after May 1968 and the liberation of morals in France to be told about particular sexual habits of the bonobo, with the goal of stirring up public interest. Nevertheless, these sexual practices are not confined to bonobos: fellatio (between males and females, or between males), face-to-face copulation, and masturbation by both females and males all occur in orangutans as well.

When Tubo Becomes More Insistent

At this time, Wattana goes through a phase that veterinarians describe as "depressive": Tubo is constantly exerting pressure on her to establish his authority and satisfy his sexual urges. She withdraws to a corner of her cage and remains prostrated, much like Nénette during Solok's reign. Tubo, on the other hand, starts developing a facial disk, which marks the second stage of development for male orangutans, who are the only anthropoid apes to have two phases of maturity (i.e., bimaturity).[32] In the first stage, males do not yet have facial disks, and are almost identical to females in general appearance, and of similar weight. Males will copulate with females, but use reproductive strategies different from those of males that have reached the second phase of development, and who have priority access to females. The second phase begins when a male is between the age of ten to fifteen for some, or fifteen to twenty for others. Muscle mass increases, as well as the hairiness of arms and back. The throat swells, and a double chin becomes apparent. In parallel, large laryngeal throat pouches develop, allowing the males to amplify the sounds they emit and produce calls that play a role both in defending territories[33] and in sexual selection. These males emit territorial *long calls*, which are very impressive, and do so several times a day.[34] These calls constitute a vocal signature, which is audible over a radius of more than half a mile. Calls are aimed simultaneously at other males, in order to avoid conflicts, as limits of territories are thus marked out clearly, and at receptive females that seek the protection of mature males. Male orangutans with facial disks become intolerant and aggressive towards their competitors. Additionally, their cheek pads develop, rendering these large males even more spectacular.[35] Sometimes, the surface of the skin of the facial disk seems to quiver. It is possible that these manifestations, which are almost imperceptible, are a signal for some emotion, yet it is difficult to offer any kind of interpretation: display of interest, slight questioning, skepticism? This disk also represents a factor that differentiates this Asian great ape from humans, who, when confronted with this face, find it more difficult to identify with orangutans. Projection-identification is easier with chimpanzees and bonobos.

Power Struggles

Tubo tests his newfound potential on his mother as well. Wattana supports him and joins forces with Tubo against the matriarch. Even so, you truly cannot teach an old dog (or in this case an old *ape*) new tricks. Nénette is quick to put her son back in his place and regain the scepter of power. In this way, great apes are constantly in search of bearings (which can only be provisional, the animal technicians saying that "with apes this changes all the time"), ceaselessly testing the strength of their alliances, hierarchies, or even the bonds between partners. Even in the case of captive groups of primates, which are limited in size, social relationships are also constantly being redefined. It is hence important for each individual to know the status of each member of his group, so as to be able to position itself within the community, and, in this way, render future interactions as predictable as possible. Living in groups is in fact especially risky as the primates are all confined in a small space with no possibility of escape. Attacks and conflicts can make a captive primate's life a real nightmare. Communal living in restricted spaces often gives rise to stress and can sometimes lead to the exclusion of particular group members. The hamadryas baboon colony at London Zoo, studied by Solly Zuckerman in the 1920s, provides an extreme example.[36] The imbalance between the numbers of males and females led to an increase in aggression and the extermination of a large part of the troop. In *Chimpanzee Politics: Power and Sex among Apes,* Frans de Waal describes how three male chimpanzees at Arnhem Zoo—Jeroen, Nikkie, and Luit—came into conflict in their attempts to take over power. Their leadership struggles culminated in the castration of Luit by the other two protagonists, who were allied against him.[37] It is customary to praise the peaceful nature of bonobos. Yet a feud between two *Pan paniscus* at Planckendael Zoo lasted for about a year. The younger of the two, and the one with more relatives present, Redy, eventually succeeded in imposing his authority on Kidogo, who lacked support from family members.[38] Hence, it is the apes or monkeys themselves that negotiate the core and overall shape of their constantly fluctuating social organization, with the key difference in captivity being that here group composition is imposed by humans. Moreover, as has been shown by numerous primatologists, among them Goodall, Lancaster, Altmann, Rowell, and Strum,[39] females play a decisive role in these arrangements.

Playful Moments

When the weather is suitably mild, the orangutans spend time in the outdoor enclosure, which is far larger than their quarters in the ape house. They can then let themselves be enveloped by the warmth of the sun, feel the breeze brush their skin, and experiment with different kinds of contact with visitors: there is mesh above the exclosure windows, which allows the exchange of sounds, as well as the transfer of objects from inside to outside.[40] Establishing a form of social communication in this way,[41] Tubo seems to derive pleasure from dragging strips of paper or cloth along the window, while waiting for a visitor to grasp one of the ends and pull on it, so that he can start a game with that person, easing off whenever necessary to rekindle interest. In this way, the objects that the keepers provide for their charges serve as means of exchange. For example, Tubo passes a small cord into the cage of his parents, and then, pulling on one of the ends, "plays" with Dayu, his young brother, who takes hold of the other end. When it is sunny, Christelle Hano sometimes gets out a hose to spray the apes with water. Wattana attempts to grasp the jet of water or catch droplets, making the keepers laugh. She paddles in the puddles that form, makes clapping noises by smacking the water surface with her hand, or sucks up the liquid before spitting it out with delight. She is sometimes allowed a paddling pool with toys. One day, on her own initiative, the adolescent filled up an inflatable balloon with water and made several holes in it, with the jets of water spraying from the balloon resembling a fountain.

It must be noted that Wattana continues to be passionately interested in the know-how of the humans who surround her: she is always on the lookout for opportunities to learn, and is able to mobilize her complete attention for extensive periods when their skills, techniques, or activities interest her. The adolescent endeavors to spend time with them, to watch closely what they get up to. She stays for long periods close to the grating, from where she tries to catch sight of their activities. In zoos, however, spaces are carefully circumscribed: on one side the humans, on the other the "beasts"; heavy metal doors or panes of glass serve as a hermetic barrier separating them.[42]

Training

Having reached adolescence, Wattana and Tubo really make life difficult for their keepers: they block the passages between two enclosures preventing the closing of the sliding gate. They spit or urinate on their keepers to attract their attention when they do not give them sufficient time or show enough interest. They devise diverse strategies to ensure they are not separated. When one of the curators of the Ménagerie, Géraldine Pothet, offered Wattana daily training sessions in April 2005, she was therefore plainly interested. Employing operant conditioning coupled with positive reinforcement as a training procedure,[43] the trainer and the orangutan are separated by a grating (in "protected contact"). These sessions, which are conducted with the help of a clicker, and which last about ten minutes each, are aimed at teaching her a sequence of gestures, so that she can be treated more easily without any need for an anesthetic. The movements she is taught have to be carried out in a predetermined order: reaching out with her arm so a blood sample can be taken, showing a specific part of her body to the curator, opening her mouth, showing or giving an object, placing both hands on the grating, etc. In a similar way, the keepers also aim to use these methods to teach her mothering skills, as they think she is "full."[44] So they gradually begin to participate in the training sessions. After a few days, Wattana succeeds in reproducing all the motions she has been taught, in the desired order, even before this is asked of her.[45] It is evident that she greatly appreciates these activities, which give her the opportunity to spend precious moments with her keepers. The fact that she was able to exercise her extraordinary cognitive and manual abilities certainly also played a part. When subjected to the same training methods, Tubo and Dayu were not nearly as efficient. Wattana's impressive performance remains unique.

Now It Is Wattana's Turn to Become a Mother

Towards the end of 2004, Wattana seems to be in poor emotional condition. In fact, she is expecting a baby. The pregnancy is confirmed by an ultrasound scan on January 19, 2005. Her keepers then start to enter her cage again, in order to prepare her for the birth, as well as for mothering. As a result, the young female rediscovers face-to-face encounters with her keepers. While these direct contacts are unproblematic when the experienced keepers (whom she has known since she was young) are involved, contacts with less seasoned keepers are riskier. The training sessions are now geared towards teaching Wattana to take care of a baby and to present it to the keepers in case of need. Yet Wattana is as carefree as ever and seems poorly prepared for becoming a mother. A second ultrasound scan is done on May 29, 2005. The pregnancy is six months along.

On August 29, 2005, she gives birth to a daughter, who is given the name "Lingga." She gives birth alone. A few minutes after the birth, she rejects the newborn. Perhaps she associates the baby with the pains endured during the contractions and the subsequent birth. Having distanced herself from the newborn, she returns towards it, curious. Lingga extends her tiny hand towards her mother, yet this gesture simply causes her to flee. Later,

Figure 2.8 Wattana and her daughter, Lingga, soon after her birth, courtesy of Christelle Hano, French National Museum of Natural History, Paris, 2005.

Wattana again attempts to approach and lies down near her daughter. The head keeper, Gérard Dousseau, has the impression that she would like to take action of some kind, but does not know what to do. Furthermore, she displays some degree of fear, and even disgust, at the sight of this noisy and bloody newborn. The keepers at the Ménagerie wait a few days before intervening. They want to give a chance to mother and child, and hope that contact will be established. Nevertheless, they are concerned about endangering Lingga, so when they see that there is no progress they decide to take the baby into their care.

Surrogate Parents

Thereafter, keepers share the duty of bottle-feeding the newborn eight times a day between 7 in the morning and 11 at night. Lingga forms really strong bonds with her "substitute parents." When she drinks from the

Figure 2.9 Lingga, photo by Chris Herzfeld, Paris, 2005.

bottle, the similarities with human babies are striking: the same respiratory patterns, the same looks, the same way of curling up in the "parental" arms. The keepers notice a strong resemblance between Lingga and her grandmother, Ralfina, an orangutan they were very familiar with. They belong to that rare breed of people who can spot characteristics associated with genealogies through physical similarities passed down from one generation to the next. The keepers persist in trying to reunite Lingga and Wattana, but without success. The female does not know how to "be a mother," having never had the opportunity to observe mothering behavior in her environment. After all, she was only four years old when Dayu, Tubo's brother, was born (a few months after the arrival of Wattana and Vandu), and primarily preoccupied with adapting to her new life. Fascinated by the games played by her conspecifics, she was undoubtedly not really interested in the behaviors Nénette displayed when interacting with Dayu. Moreover, merely *observing* this type of behavior could never replace personally experiencing a relationship with a mother. Her inability is further compounded by her immaturity. Like Ralfina, Wattana was not even ten years old at the time of the birth. Although she does exhibit some curiosity when looking at her daughter (much like being exposed to some new toy), she never tries to hold her. In her eyes, the baby also seems to be a competitor, demanding more than its fair share of the attention of the surrounding humans.

Lingga

Reared by her keepers, Wattana's daughter grows up. She explores the enclosure she has been placed in, first on all fours, and then later standing up on her frail legs, with her hands gripping the bars of her playpen. Inquisitive by nature, she observes her elders through the window and then returns to her activities. When the keepers are absent, Lingga finds ways to keep herself occupied. She's never short of ideas, always able to invent new and "important" tasks for herself, such as reaching a toy placed high up in her cage, climbing to the top of a wooden tower, dismantling the multicolored objects that are placed in her enclosure. Later, several attempts are made to integrate her into a small group consisting of Nénette and Dayu. Unfortunately, Nénette, the matriarch, attacks her. Moreover, after playing with her initially, her son is also rough with Lingga. So the young ape returns to her cage, where she stays until her departure for Monkey World Ape Rescue Centre in Wareham, United Kingdom, on July 22, 2007.

3 Living at the Zoo

Deprived of a space, no living being can live.
—MICHEL SERRES, *Le Mal propre*, 102

At the time when the first orangutans arrived in Europe, scholars had only a very vague idea of their lifestyle in the natural habitat. They merely had access to rather limited descriptions and engravings,[1] and were wary of the (presumed fanciful) tales of long-distance travelers in which these "men of the forest"[2] were described. This wording is a perfect translation for the bewilderment felt by naturalists, who were uncertain of the status they should attribute to this creature that resembles humans yet lives like an animal. They were, however, soon able to put their knowledge to the test while making the first direct observations of the limited number of orangutans that made their way to the West, sometimes (but very rarely) alive, more often as skeletons or preserved in alcohol. Here are a few famous examples.

Humans and Orangutans: First Encounters

In 1776, Arnout Vosmaer, director of the cabinets of curiosities and the Ménagerie of Prince William V, of Orange-Nassau at the "Kleine Loo" (The Hague, Netherlands), had the opportunity to observe the behaviors of a female orangutan, undoubtedly the first to arrive alive in Europe.[3] The she-ape enjoys the company of humans. She is able to move about bipedally, appreciates fine wines, and eats strawberries one-by-one with a fork. She is also capable of untying the most complicated knots, cleaning the floor and the furniture in her room with a rag, and using napkins and toothpicks. She proves to be very inventive, using a nail as a lever and fashioning herself a pillow. Vosmaer is utterly baffled by the young female's behavior. After selecting a place for her bed, she spreads out a sheet on the floorboards, gathers up a pile of hay at its center, and folds up the four corners of the cloth. She then carries this pillow, with great care, to her bed, and lays her head on it after pulling up the blanket over herself. In 1795, Georges Cuvier and Étienne Geoffroy Saint-Hilaire wrote their *Histoire Naturelle des Orangs-outans* (A Natural History of Orangutans),[4] based on the skeleton of a Grand Pongo de Batavia or Grand Pongo de Wurmb (*Pongo wurmbii*) that they had requisitioned in the collections of the prince of Orange.[5] It is the first time that they really show an interest in apes. Their driving motivation is actually ideological: they want to discredit any notion of species intermediate between man and beast, and to reaffirm the preeminence of humans, just when others were defending the idea of a simian origin of humans, or even wondering if great apes might possibly be part of humanity.

Yet, the ape they studied was just over four feet tall,[6] almost the height of a human. Furthermore, this primate had been described in a scientific report, written in 1780 by Baron Fredrik von Wurmb and published by a scholarly society. So, this was not a description based on travelers' tales or reports of maritime captains, accounts that had long been viewed with suspicion. Previously, the rare ape specimens of that arrived in Europe had all been very young individuals. Naturalists could differentiate these from humans with ease due to their small size, and hence maintain a strict separation between humans and apes. In fact, the Pongo de Wurmb is the first *adult* orangutan (according to current criteria) to be observed by European scholars, and as a consequence, they considered this specimen to belong to a new species: they could not imagine that an Orang-outan (according to contemporary opinion) could reach a height of more than thirty inches. At that time, there was still confusion regarding differentiation of chimpanzees and orangutans, as well as their developmental stages, species, and subspecies. It was not known that the chimpanzee (the "*Jocko*") was a native of Africa, while the orangutan (the "*Pongo*") was from Asia. The names *Jocko* and *Pongo* were used interchangeably and haphazardly.[7] After having been called *pygmies* in the seventeenth century,[8] apes were then grouped under the generic name *orangutan* in the eighteenth century, increasing confusion even more. Moreover, scholars recognized multiple varieties: Orang-outan, Pongo, and Grand Pongo de Wurmb, without realizing that these were the same species, and simply individuals of different ages. Classification of primates was then governed by a single criterion: the facial angle defined by Camper, which expresses the degree of prognathism, the projection of the "snout." The closer this angle resembles the ideal beauty of Greek statues (i.e., 90 degrees), the more it represents an index of superiority. In order to demonstrate that the Grand Pongo de Batavia (actually an adult orangutan) was "no more than an ape," and to remove him as far as possible from humans, Cuvier and Geoffroy Saint-Hilaire relegate the species to the bottom rung of the primate ladder,[9] among the baboons. The specimen's facial angle seems to be limited to 30 degrees, while that of the *Orang* (also an orangutan, but much younger) was 60 degrees. So, on the primate scale, the species is demoted from first place, attributed to it by Blumenbach in 1782, to sixth place. This makes it difficult to defend the idea of a common origin. A threat to the preeminent position of human beings, because of its large size and marked resemblance to humans, the orangutan would stay downgraded until 1835, the year when Richard Owen, a biologist and specialist of comparative anatomy, would elevate the status of the species, placing it in second position, right after the chimpanzee.

Imperial Ape, Knowledgeable Ape, and Talking Ape

In 1808, General Decaen, Governor General of the French East Indies, brought back a treasure as part of his cargo: a young orangutan, the first to have made it to France alive. He presented the ape to the Empress Joséphine. She had in fact recreated her own Petit Trianon[10] at the Château de Malmaison, her usual residence, and had equipped the park with a model farm, as well as a menagerie for exotic animals, which began to compete with the Ménagerie of the Jardin des Plantes after 1800. When a rare species is donated to the government, it is difficult to decide where it should be housed. As a notable example, Joséphine demanded that two kangaroos impatiently expected by the professors at the Natural History Museum should be handed over to her. Because France had no colonies in Australia, marsupials were lacking from French collections for some time. On the other hand, it is also true that the sovereign donates some specimens to the Muséum. But she keeps the baby orangutan, which is most probably about eleven months old. Having been around humans from her earliest infancy, the young female settles into her new life very easily. She is dressed in a frock coat, dines at the table with the Empress, acquires refined manners, spends time with the Emperor, and sleeps in a bed, with a blanket pulled up over her chin. She is very sociable and likes to share kisses and cuddles.[11] Frédéric Cuvier notes that she knows exactly how to manipulate the people around her. During fits of feigned anger, she observes the effect her cries are having, and then resumes her vocal protests when she does not get what she wants. She is also able to understand the feelings of her human companions and attribute feelings to others. However, she must have been very young indeed, as it was reported that she died at the age of eighteen months, afflicted by "inflammation of the intestines." London Zoo opened its doors to the public in 1828. It received its first chimpanzee, named Tommy, in 1835. Later, in 1837, the first orangutan arrived on the scene. As a result, Charles Darwin finally had an opportunity to observe a living representative of the species. One of her conspecifics, Jenny II, would later be presented to Queen Victoria, who was fascinated by the delicate way in which she would drink tea, yet at the same time appalled that she was so "painfully and disagreeably human." At the end of the nineteenth century, William Hornaday, the future director of the zoo in New York, spends time with a young male orangutan in Borneo. The ape, Dohong, is attached to Hornaday as if to a father. He effortlessly learns all the ins and outs of a human's daily life.[12] At the beginning of the next century, Hornaday describes the exploits of two orangutans at the New York zoo in his study on the intelligence of wild animals. He declares that the representatives of this Asian species are "naturally docile and affectionate," and more serene and "philosophical" than chimpanzees. He explains that they can easily be taught to dress or undress themselves, sit at the table, and eat with cutlery like humans.[13] He further relates that the orangutan Rajah, an excellent imitator trained by the zoo keepers, was able without difficulty to drink tea, use keys and even try out every key in a bunch before finding the right one, hammer in nails, use a toothpick, or spit. He was able to ride a bicycle after only three lessons. The keepers also taught him to sit at the table, exhibiting sophisticated manners, within two days. In addition, he knew how to light and smoke cigarettes. Dohong continued to develop his faculties of invention and reasoning, and further demonstrated surprising mechanical skills. He discovered (or invented) the principle of leverage and applied it to a variety of situations. Another orangutan at the zoo was specialized in braiding ropes from a variety of materials, and tying them to structures within his cage, so as to swing from them. Hornaday insists that there are major differences between individuals. Chantek, who was born at the Yerkes National Primate Research Center (at Emory University, Atlanta) on December 17, 1977, would also have his moment of fame. He belonged to that select community of apes that learned to use sign language in the 1960s to 1980s. Chantek was transferred to Chattanooga in October 1978, at the age of about nine months, and then raised like a human child by the anthropologist Lyn Miles for more than seven years. He learned the language of so-called deaf-mutes and established real dialogues with Miles. One evening, when they were together on the hills near the University of Tennessee, Chantek used signs to ask her "What is that?" while pointing at the moon, which seemed particularly large and bright that night. According to Miles, he was wondering about the celestial body and shared his puzzlement with her.[14]

Men of the Forest in Captivity

According to the international *Studbook* for the species, approximately 3,000 orangutans have lived in captivity since the beginning of the twentieth century.[15] Two hundred fifteen orangutans are currently living in 55 zoos accredited by the Association of Zoos and Aquariums, in Canada, the United States, and Mexico.[16] In 2010, there were about 339 orangutans housed in 70 zoos across Europe.[17] At the beginning of the twentieth century, the first four hundred captive orangutans were almost exclusively individuals that had been taken from the wild. In the 1960s, individuals were still captured in the wild, but births in captivity had become more common. In 1988, 70% of individuals were zoo-born, rising to 90% by 2010.[18] Since 1989, the captive orangutan population has been stable, with births compensating for deaths. The first female mentioned in the *International Studbook* was called Cleo (number 1). Captured in 1916, she resided at the Berlin Zoological Garden from May 1926 to November 1943. Number 2 was the male Guas. Like the female at number 3, Guarina, he lived at Philadelphia Zoo from 1919. They died in 1977 and 1976 respectively, at 60 and 59 years of age, bearing witness to exceptional longevity. After their capture in Sumatra, in 1927, they had first been housed at Rosalia Abreu's colony in Havana. Abreu was fascinated by great apes, and had developed a specific care regime and very strict rules of hygiene, at a time when no one knew how to look after them. Anumà, the first chimpanzee to be born in captivity, was delivered at the colony on April 27, 1915.[19] The founding father of American primatology, Robert M. Yerkes, in fact traveled to Cuba to consult Abreu, who had become an international expert in the matter. It seems that her know-how was transmitted to the keepers at Philadelphia Zoo, to the great benefit of the primates living there. On January 22, 1935, Guarina gave birth to the little Cinderella (number 8). This was the first birth of an orangutan in a zoo and was therefore a major event. The pair would later have a second infant, Ivy (number 11), born on June 13, 1937. Guarina went on to give birth to another seven babies.

Apes among Humans

By becoming residents in zoos, orangutans progressively entered the worlds of humans, very different from those they had inhabited prior to their capture. Like bonobos, chimpanzees, and gorillas (their African cousins), orangutans live in tropical rainforests. Apes are tightly linked to the forest environment and have developed a true complicity, or even a "connivance," with all its components. Every tree, bush, leaf, liana, branch, trunk, twig, fruit, or berry bears multiple meanings and is embedded in a network of relationships that is beyond our understanding. What can an orangutan, who is adapted to arboreal lifestyle, achieve in a totally different world, enriched with new features, yet still impoverished compared with the abundance of the forests? Having colonized their forest habitats so successfully, how can apes manage to live in radically different spaces such as cages in the zoos of the West? In order to exist, all living beings must take possession of the places they inhabit.

As far as mammals are concerned, the intimate relationship between the environment and the individual already starts to develop within the mother's body. The fetus is surrounded, enclosed, in a warm, secure place. The individual is already cradled by particular rhythms and is exposed to sounds, notably the heartbeats of the mother carrying it. After it has been born, the young one gradually establishes relationships with its environment, at first consisting of the mother's body, and her fur, where the infant finds refuge. This contact is extended in the nest that it shares with its mother for several years, a nest that Bernstein describes as "a circle around a body."[20] The young primate thus weaves a "solid link . . . between a body and its nest, between a life and its location."[21] Enclosing the individual rather like pregnancy, this nest, which is characteristic of all apes, becomes the vector for a "return to the womb," and becomes a place to sleep, play, have physical contact with the mother or some other conspecific, seek a little peace and quiet, protect itself, hide, reproduce, look after itself, suffer, or . . . die.[22] Chimpanzees and bonobos spend almost half their lives in their nests. According to Egenter (1990), great apes construct 10,000 to 15,000 nests during their lifetimes, while Fruth and Hohmann (1994) report that chimpanzees build around 19,000.

Being "At Home"

As mammals that have been influenced by a long childhood spent mainly close to their mothers, both humans and great apes require confinement, warmth, and intimacy: "The need to keep the body in multiple contact with 'partitions' is apparent in most mammals, especially those that are territorial."[23] This requirement, which is partly determined by mammalian physiology (notably because of the need to maintain constant body temperature), underlines the contrast between the intimate sphere, the interior space, the nest, and the exterior sphere, the bush, the forest. Whether the animal is eating or resting, this "creature on the lookout" of Gilles Deleuze is constantly on its guard. The animal is never quiet, always looking back over its shoulder.[24] In this way, the individual marks out a limit between a *home* that is sheltered from predators, and an *exterior* that is open to all forms of danger. This limit is directly linked to the theme of territorialization discussed by Deleuze and Guattari[25]: "But home does not preexist: it was necessary to draw a circle around that uncertain and fragile center, to organize a limited space."[26] Each mammal thus creates for itself an intimate space where it can find warmth and safety. This primal need to mark off a territory is just as essential as drinking, eating, establishing social relationships, or copulating. How do captive orangutans create their own space in the enclosed areas that are imposed on them? How do they fashion their own individual refuge while they are living in captivity? In fact, even if their boundaries are put in place by human beings, primates will superimpose their own marks, even going so far as *enclosing* the cage to which they are assigned. Two examples testify to this. The first, related by Gérard Dousseau, involves Toto the gorilla. One morning, the head keeper found the ape lying down in the middle of his platform. As he showed no response to the arrival of his keeper, Dousseau entered the cage and approached the gorilla, despite recent "aggressive" episodes and expressions of rivalry. Toto stood up suddenly, grabbed the keeper by his legs and lifted him, seemingly without effort. He then deposited the keeper outside the cage and closed the door angrily, before returning to lie down. In the second example, Marie-Claude Bomsel (then director of the Ménagerie) entered Nénette's cage with a photographer who hoped to take a few exceptional photos. At the time, people still occasionally entered her enclosure, but only as long as either G. Dousseau or M.-C. Bomsel, her adoptive parents in a way, were present. The two visitors approached her, venturing ever closer. Yet it seems that they came a little too close for the orangutan's liking. An exasperated Nénette charged her visitors. The two very obviously unwanted guests took to their heels and rushed towards the exit with the female in hot pursuit. In their haste, they "forgot" to close the cage. One might imagine that Nénette, presumably a "freedom-loving" being, would have seized this chance to finally take a stroll in the open air. But this, as it turned out, was not at all the case: like Toto, she slammed the door with full force behind her unwelcome visitors, and reclaimed her "home," perhaps limited, but still a "home" of her own.[27] She also made no attempt to get away.[28] Despite omnipresent and burdensome boundaries, with no possibility of expansion, captive great apes themselves enclose and delimit their living space. This occurs although they are not under threat from predators. What is at stake is the necessity to have a "place of their own" at their disposal. In a zoo, territories cannot be demarcated for purposes of foraging or defense strategies (and hence with a functional context), but they are well and truly places where individuals try to fashion an *existence* for themselves, despite the restrictions of the "dispositive," or the apparatus, in Michel Foucault's sense.[29]

"I Belong to a Space or That Place Belongs to Me."[30]

A multiple array of disparate links is thus created between the great apes and the space that they occupy. At the Ménagerie, for instance, the apes prefer specific and precise locations within their enclosures. Nénette always builds her nest in cage number 2, at the top and to the right. At the time when Solok was her companion, she would stay all day on the mezzanine, hiding in a cardboard box. The apes at the Ménagerie do not have access in either the interior or exterior cages to "withdrawal zones," spaces where they can hide from public view. The two possible strategies to avoid being in the public gaze all day long are hence either to seek refuge higher up, at the back of the mezzanines, or to hide in some kind of container, such as a basin or large cardboard box. This is what Nénette systematically did. Visitors would sometimes catch a glimpse of a hand, or a foot, sticking out of this improvised shelter. Wattana, on the other hand, actively sought contact with people. She would frequently squat in an enormous drum with its opening facing the window, the "visitor's side." Tubo, in turn, would occupy his enclosure by arranging it to his liking, in an extremely precise and even obsessive manner.[31] When the young Tamu (who arrived in November 2007, accompanied by her mother Theodora) made a habit of taking down the tire that the adult male had carefully suspended high up, Tubo would hasten to put it back in its place. In another example, as soon as they were able to find a sufficiently hard tool, the apes at the Ménagerie engraved lines on the metallic surfaces in their enclosures, leaving their marks in the form of intermingled lines. They would also tear away the rubber surrounds of the glass panes, and deposit excretions and gobbets of spit at various locations.[32] These modifications all serve as signatures marking out their domains. The keepers know better than anyone that apes *inhabit* space in a multitude of ways: cleaning of the windows with their urine; discreet drumming against the walls; noises that are peculiar to each individual; hands that are extended through the railings; exchanged glances; moments of silence; "home-made" productions or installations in the enclosures—for example, jute tents hung from beam to beam, paper strips draped across the whole cage, or tied to particular structures: wire mesh, rings, or the ceiling gridwork. In this way, they occupy the places that are imposed on them by amplifying presence in these confined quarters. These declarations of their presence do not seem to mean "This is mine,"[33] as would be the case in humans, but rather: "This is my home."[34] Beyond the need to create a home area, at the same time captive orangutans engage in the key issue of a *territory*, which is marked out as an efficient means of avoiding conflicts.

Ape parents transmit this manner of occupying spaces to their offspring. Although these zoo residents show much goodwill and try hard to establish a home in the spaces assigned to them, they are rarely allowed to stay there for the rest of their lives. Nénette is an exception to the rule; she arrived in Paris at the age of three and never left the Ménagerie. But most primates are transferred from zoo to zoo, in compliance with "breeding plans" regulated by international associations of zoological gardens. They are hence frequently separated from those humans or conspecifics with whom they have patiently built relationships of trust. These, too, represent important components of the "homes" that apes construct for themselves. Following each transfer, the apes therefore have to form relationships with their new enclosure companions, get to know their keepers, and rebuild an existence in the places that are imposed on them in such a high-handed manner.

Intermingled Socialities

According to the keepers at the Ménagerie, the orangutans are utterly bored on the rare days when no visitors are present because of closure or strikes.[35] In fact, they spend a very large part of their day carefully observing members of the public that parade through the ape house. If Tubo saw somebody to his liking approach the enclosure, he would press his face up against the window. When he recognized someone, he would move his lips, as if blowing a kiss. The animal technicians also provide crucial "environmental enrichment" for captive primates. They contribute to enhancement of the apes' social life, which is severely impoverished compared with their natural habitat, where encounters with conspecifics or members of different species, both animals and plants, are infinitely more numerous. Keepers and apes show a mutual interest in one another. The former adopt the role of, somewhat peculiar, social partners for the apes, while the latter focus their attention on the keepers. Great apes often find ways of attracting additional attention. For example, the orangutans at the Ménagerie unscrewed a large number of the bolts in their cage and hid them all. The keepers were worried about the foolish things the apes could use these metal pieces for, notably for scratching the windows, so they tried to negotiate their return

Figure 3.1 Wattana in the arms of her keeper, Frank Simonnet, in the monkey house's backstage, courtesy of Frank Simonnet, French National Museum of Natural History, Paris, 1998.

using cookies. The apes very quickly grasped the concept of bartering and sought to reap maximum profits from this transaction, handing over the bolts one-by-one so as to receive a greater number of treats. Visitors, keepers, and animal technicians represent essential components of the world of captive great apes, which observe them as much as they are observed by them. Drawing upon these long hours of observation, the apes attempt to replicate, and to adopt, various behaviors and skills. Some humans become models and *charismatic leaders* for them: they are not only scrutinized for what they *do*, but also for what they *are*, one of their characteristics being to immediately command respect from the rest of the community.[36] As noted by the scientific journalist Eugene Linden, it is important that animals are able to attach themselves to their world in a constructive manner. They thus show considerable goodwill in collaborating with humans.

Behavioral Plasticity and Flexibility in Habits

When living in close proximity with humans, great apes demonstrate surprising plasticity, exceeding what one might imagine them capable of. They adopt human habits, copy their actions, and take possession of their objects. For example, the orangutans at the Ménagerie drink tea every afternoon. They prefer to eat their yogurt with a spoon. Chantek, the talking ape already mentioned at the beginning of this chapter, in fact has the habit of visiting the toilet. When he returned to the Yerkes National Primate Research Center after having lived together with Miles for nine years, he was distraught, not knowing how to cope in this new universe deprived of conveniences. The female chimpanzee Linda arrived at Madrid Zoo in 1992. Having had her teeth removed by her previous owner, she invented a culinary technique for enjoying her favorite food: she would scrape fruits against the rough wall of the inside enclosure and lick the pulp that she extracted in this way. This operation to reduce food items to juice or mush demands a lot of concentration and can last anything from tens of seconds to roughly fifteen minutes. The other apes progressively began to imitate Linda, and applied her technique to the apples, carrots, oranges, mandarins, tomatoes, cooked potatoes, and lemons they were given. Linda was thus the originator of a novel kind of community know-how. According to an article published in *New Scientist*,[37] this was the first time that such preparation of food items, or that kind of transmission of a technique, had been observed. Samuel Fernandez-Carriba, a primatologist at the Universidad Autonoma de Madrid, uses the expression "culturally transmitted customs."[38]

Bipedalisms

In principle, apes are *quadrupeds*. Yet, it is notable that, stimulated by human beings, some captive individuals adopt a form of locomotion differing from the form that is seemingly strictly determined by nature. Raised by humans at the Bronx Zoo, and now at the Zoo of Louisville in Kentucky, the female gorilla Paki moves around almost exclusively bipedally. She seems to be totally comfortable standing up with her chest perfectly upright. The male orangutan Teak, who lives in the same zoo, also employs this mode of locomotion. He is not an occasional biped, but moves around in this way most of the time. Two other apes, the chimpanzees Oliver and Poco, are also known for the same bipedal capacity. Oliver, who had been raised by Frank and Janet Berger, died in a Texan sanctuary (Primarily Primates, San Antonio) in June 2012, at the age of approximately 55. Like many apes that have grown up within human families, he preferred women to female conspecifics, and refused the company of other chimpanzees. He also loved to paint and in fact had so many human attributes that for a long time it was thought that he was a human-chimpanzee hybrid, a possibility that was later disproved using genetic tests. Oliver served as a model for the latest movie with the theme of a planet ruled by apes (*Rise of the Planet of the Apes*, 2011). As for the male Poco, he lived for nine years alone in a cage, waiting to entertain potential clients. This cage was so small that he was obliged to stand upright. By the time he moved on, he had therefore acquired the habit of being in a bipedal position. When he was admitted to the Sweetwaters Chimpanzee Sanctuary in Kenya, he was finally able to enjoy a decent life, yet continued to move around on his hind legs. He can even run over short distances, as documented by a video.[39] Moreover, it would seem that he passed this mode of locomotion on to the youngest members of the group. Like Poco, the female orangutan Mari, resident at the Center for Great Apes in Wauchula, is fully bipedal. Having lost both her arms at a very young age, she learned to move about in an upright position. She is even able to move around using the sanctuary's wire mesh tunnels that link enclosures, even though these tunnels involve rather abrupt declines and slopes. Another female chimpanzee, Annie, who is 38, lives at the Chimpanzee Sanctuary Northwest, in Cle Elum (Washington state). She, too, is frequently observed standing up, in the tall grass. Now housed at the zoo in El Paso (Texas), the Sumatran orangutan Butch (born on August 19, 1985) has also adopted bipedal locomotion. Furthermore, bonobos are well known for their ability to practice this mode of locomotion. Rather than embracing the idea of linear evolution from quadrupedalism to bipedalism, Pascal Picq explains that all apes possess a preexisting capacity[40] for bipedal locomotion. Hanging from and moving through trees, advancing using brachiation and ascending along tree trunks, in a vertical position, are all actually included in the locomotor repertoire of apes and prime them for standing upright. Accordingly, Picq demonstrates that there are different forms of bipedal locomotion and prefers to refer to *bipedalisms* in the plural. The swaying gait of australopithecines, who swiveled their hips and shoulders and were unable to run, was very different from that of *Homo ergaster*, an accomplished biped with the capacity to run and walk over long distances. This type of bipedalism is seemingly a precursor of bipedalism in *Homo sapiens*. It is often argued that the capacity for such a mode of locomotion is better developed in the species that are genetically closer to humans (i.e., bonobos and chimpanzees). Yet the cases of these apes that live in close proximity to humans and move around in the upright position attest to the fact that all five anthropoid ape species practice bipedal locomotion, of different types, depending on their proclivities and physical attributes. All species share this tendency to varying degrees. It is thus important not to dwell exclusively on the question of the origin of bipedalism, but also to consider how this ability arises in a developmental sense. The fact that primates that are closely related to humans adopt bipedalism obliges us to take into account a hitherto neglected dimension, that of sociocultural imitation. As they are raised *among* (or even *by*) humans, some apes quite logically mimic the mode of locomotion of those that surround them. In the environment peculiar to zoos, humans, who are all bipedal, are present in far greater numbers than quadrupedal primates. This environment, the models embodied by human beings acting as *charismatic leaders*, their encouragement, and the strength of the bonds that unite them, prompt some captive apes to stand erect. In addition, the bipedal position allows face-to-face encounters, as well as richer interactions, with humans. These various elements notwithstanding, we continue to ignore the importance of the social dimension, the weight of culture, and the power of *emulation*, in driving the emergence of particular behaviors, as soon as one enters the realms of "animality" and "nature." When presented in science lessons, wedged in between nutrition and reproduction, locomotion is one of the traits that is thought to be strictly biological. However, although rarely discussed, these compliant apes exhibit a certain degree of freedom compared with what appeared to be hard-wired in biology or instinct. Such locomotor plasticity is further supported by skeletal plasticity, which is adapted to the biomechanical constraints associated with bipedal locomotion.

Cultural Heritage and Adaptability

Whether they are cooks, handymen, tea drinkers, or bipeds, these compliant great apes, which I would call "conciliatory apes,"[41] display their capacity to take on particular typically human skills and multiple ways of living. Moreover, because of this adaptability, apes that have spent a long time around humans are far more difficult to rehabilitate than other species. Indeed, this has meant that most attempts to release ex-residents of zoos into the wild were doomed from the outset. These great apes do not possess the crucial knowledge and culture required for survival in the forest. The world of the zoo is all that they know.[42] Besides, adults are less adaptable than youngsters. With careful preparation, some apes living in African or Asian sanctuaries have nevertheless been successfully reintroduced into their natural habitat. The rehabilitation program set up by Claudine André in the Democratic Republic of the Congo provides an excellent example. The success of this project has been notably facilitated by the particular intelligence and sociality of the species. The bonobos of the Lola ya Bonobo center were accepted by conspecifics already living in the release zone. What is more, cooperation with the local people had been officially negotiated, so they were actively involved in the project. In fact, since the 1980s, it had been fully understood that animals could not be protected in the wild unless local inhabitants saw a real economic benefit in doing so.[43] People working on the ground hence aimed to reconcile development strategies and nature conservation within multilateral cooperation programs. Moreover, Claudine André's bonobos had learned—guided by the most experienced members of the troop—to select fruits and other plant foods in the wooded areas of their former sanctuary. They were also accustomed to the regional produce, supplied by market gardeners living around the Lola ya Bonobo. Additionally, their keepers had complemented this learning process by making them discover the resources peculiar to their site for rehabilitation.

Denatured Apes?

Above all, zoos aim to present "wild" animals to the public in naturalistic surroundings. The bonobos at Planckendael Zoo (Malines, Belgium) are accordingly exhibited in an African village, with a jeep bogged down in a nearby river, a dugout canoe, and a village school, where the pupils' songs can be heard. Everything is set up to present the animals in a context that is as "natural" as possible. We find ourselves in Africa. This kind of staging is taken to the extreme in some American zoos. When visitors enter the "Myombé Reserve" of the Busch Gardens (one of the largest zoos in North America, located in Tampa, Florida), they are greeted by lush vegetation, morning mists, several waterfalls and exotic trees, almost outdoing nature.[44] Behind a waterfall, one can catch sight of a chimpanzee or a gorilla taking care of its young. The Florida climate contributes to the African atmosphere. The primates are staged in a way that they seem to have come to us accompanied by their natural environment.[45] Yet, in the evenings, away from the public eye, the great apes at the Busch Gardens watch cartoons or video documentaries.[46]

What is the actual status of these captive apes that are presented as wild animals, yet nevertheless serve as instruments in our economic, cultural, and social activities? As creatures of the borderlands and inhabitants of transitional zones, imported from their tropical forests to be housed in zoological gardens, they are unclassifiable, and this is what makes them problematic. In fact, these primates challenge major categorical distinctions (nature versus culture, wild versus domesticated, human versus animal), despite our conscious efforts to display them as "wild" animals. Some consider that, in a way, these primates are *denatured apes*, as well as *false apes*, as opposed to *natural apes*, or *true apes*. However, attempts to constrain great apes to these overly strict binary categories, opposed and polarized, are nullified by their extraordinary capacity to drift from one world to the other when they live among humans. The zoo environment thus renders some of their essential qualities visible: great behavioral fluidity and a remarkable plasticity in ethos. These are characteristics shared by all members of the hominoid group, to which human beings also belong. Even if some behaviors are generally stable (because they are appropriate to a given world, and a given moment), they are still not fixed: apes do not lock themselves into stereotypic behaviors dictated by "instinct." In zoos, they are propelled into worlds that force them to mobilize their flexibility, at least to the same degree as in their natural habitat, a relatively stable environment in which they have lived for millions of years. They then adopt human habits and skills, respond in inventive ways to our prodding, and enter our cultures in a most impressive fashion. Even so, this does not make them less "ape." On the contrary, they exhibit a fundamental trait of hominoids: plasticity.

Conciliation

Some describe those apes that cohabit with humans, and live in close proximity with them, as being *bi-cultural* (S. Savage-Rumbaugh), or as belonging to *hybrid communities* (D. Lestel). These views are inadequate as they are products of a dualistic mind-set that demands a radical separation between "human culture" and "great ape nature." Being *ape* or *human* does not correspond to monolithic natures that can be captured by definitive characterizations. There are no closed and definite "simian" or "human" worlds, to which external elements are attached, and that culminate in the emergence of a being split between two "cultures," oscillating between the two worlds, half "ape," half "human." I prefer to use the phrase *"conciliatory" apes*. As a matter of fact, apes integrate human properties such as habits, social activities, ways of life, and even lineaments of ethos. They extract what is needed and makes sense for them from the human structures, opportunities, behaviors, and objects to which they are actively linked. Components of their environment represent for them an array of resources, possibilities for innovation, and opportunities for transformation. It is hence not just a matter of being exposed to our influence, or of hybridization between two cultures, but rather reflects the elaboration of a *world*,[47] drawing from the ingredients available in a given place, that itself is linked to particular, and constantly fluctuating, activities. During this process, apes *appropriate* various constituent parts of the worlds peculiar to humans. Successive appropriations gradually allow the emergence of preferences, a way of being, a state of mind, a *habitus*,[48] which little by little, becomes a *foundation*. This process is thus not limited to anchoring habits. The properties that are refreshed or emerge truly become a *foundation*, the hard core, specific to a particular individuality. These properties are tightly linked to a social, cultural, and technical capital. Far from being anecdotal, such properties allow great apes to live in worlds that differ from those they have known for millions of years. Each individual thus constructs a *world*, and invents itself, in a singularizing process.

Impoverishment of the Environment

In this way, captive apes have the opportunity to revitalize latent properties that are specific to all anthropoid primates. Even so, they do this within the constraint imposed on them by their structural characteristics (anatomical, biomechanical, cognitive, social, and sensorial), on their own terms and expressing a potential that allows or excludes certain possibilities. The *plasticity* of great apes thus enables them to "construct a world" within the zoo environment, which is obviously shaped to a greater extent by *human elements* (diet, architecture, medicine, leisure activities, social partners, objects, etc.) than the forests from which they came (although these elements are not totally absent there). Nevertheless, conciliatory apes are confronted with a significant impoverishment of their habitat, as regards the variety of species, the occurrence of encounters, tastes and flavors, diversity of the landscape and resources, and social relationships.[49] Moreover, during the extended cold seasons in Northern countries they are often confined in restricted spaces with no access to the outside.

Yet, when opportunities or necessities to employ certain capacities are lacking,[50] we see losses in skills, contraction of learned activities, and greater rigidity of behaviors (and hence a loss of flexibility). This is also the case for other species. For example, if the environment of a male weaver bird does not provide the necessary materials for building his nest during the period appropriate for learning, he will never manage to build one, even when opportunities to do so arise later on.

Inventing Opportunities

So it is important that possibilities offered by humans should compensate for impoverishment of an ape's environment. The appeal of novelty for apes seems to be magnified in captivity. This is linked to a crucial neotenic trait[51]: curiosity. Keepers report that great apes, which are extremely attentive to the slightest change, notice all modifications, no matter how small. Nothing is missed by these talented observers: a new haircut or hair color, a novel accessory, a change in perfume, injuries and bandages, presence of a beard after a prolonged absence. In a way, their strong attraction to new activities or objects constitutes a kind of lifeline, a very healthy mechanism that undoubtedly allows great apes to limit possible losses in proficiency. It seems to drive them to compensate for deficiencies and losses brought about by confinement. For this process to be triggered, though, it is essential that the environment should offer at least a basic level of interesting opportunities and a certain quality of life. The future of each individual is thus tightly bound to the components available in its habitat. Accordingly, we cannot understand great apes if we do not view them as being embedded within a well-defined system. Moreover, apes invent their own opportunities for experimenting with new situations. An example of this is provided by Zomi, who was a young female bonobo when I observed the group at the Zoological Garden of Planckendael, wandering around the large interior hall with a paper bag on her head. In this way, the small female explored her environment using new methods, probing her ability to move around blind, and testing her mental representation of the area. Indeed, this behavior is not limited to great apes: at the Ménagerie, a De Brazza's monkey occasionally wandered around with a large leaf over its

Figure 3.2 Wattana, 10 years old, playing with Lego, courtesy of Christelle Hano, French National Museum of Natural History, Paris, 2006.

face, as if holding a mask. The keepers checked and found that the leaf was completely opaque. Primates thus find new challenges to face and render some situations more complicated so as to make them more interesting. Drawing solutions from the human behaviors that they observe, they are also able to improve strategies or invent new ones to respond to unusual situations. Great apes never cease to surprise us with their inventiveness. Wattana, for instance, likes to swill Coca-Cola (something she really loves) very quickly around in her mouth, to derive maximum enjoyment from its fizzy effects. She is interested in particular species of rodents or insects, and sometimes spends a long time playing with them. According to her keepers, she sometimes also places worms on her body and walks about with them with "great pride." If she is interrupted during this solitary naturalistic occupation, she will wrap up the worms in pieces of paper to save them for later.

"They Should Be Stimulated to Do What They Feel Most Inclined to Do."[52]

Human beings have responsibilities towards the primates that they have so arrogantly imprisoned. One of them is providing the means to engage in activities that interest them, making diverse opportunities available while taking their preferences into account and allowing them to do what they are most inclined to do. These activities include the whole range of occupations linked to the technological advances that are our strength. There are numerous possibilities: touch screens, joystick games, television, electronic games, computers, and tablets. Great apes are truly fascinated by these technical devices. They love playing with joystick games and very quickly understand how to handle them. The chimpanzee Ayumu (Primate Research Institute, near Kyoto) demonstrated the ability to sort numbers on a computer screen more rapidly than human counterparts. The association Orangutan Outreach, which is directed by Richard Zimmerman and fights to protect orangutans in their natural habitat, initiated Apps for Apes, a project aimed at improving the living conditions of captive orangutans by enabling them to use tablets.[53] The apes' interest in the iPad was first tested in a zoo in Milwaukee. Then, tablets were distributed in different zoos in Houston, Miami, Wauchula, Atlanta, Memphis, and Toronto. As a result, the apes are now able to draw, play, compose pieces of music, watch documentaries, or see *in real time* their former social partners, now residing in other zoos. At the Center for Great Apes in Wauchula, Pongo particularly appreciates the game Piano Virtuoso, which enables him to compose a sequence of notes on a digital keyboard, while Mari prefers the game Buddy Doodle.[54] Thanks to this application, the anthropoids can "paint" with their fingers on the tablet's touch screen. They have the possibility to select colors, and the type of line ("chalk", "paintbrush", "stump"), as well as embellish their creations with small pictures (flowers, insects, animals, hearts, and various emoticons). The case of Mari is particularly interesting, as she found a way of using the device despite having lost both arms during childhood: she drew with her feet. Patti Ragan, the founder of the center, feels that the iPads represent an excellent form of enrichment. Indeed, great apes are in constant need of new occupations. Tablets stimulate their creativity and allow them to combat boredom. Furthermore, Zimmerman showed some photographs on her iPhone to the female Sumatran orangutan, Jahe. The female seemed to recognize her relatives at the zoo in Toronto, where she was born in 1997, and which she left in 2010 when she was moved to Memphis Zoo. Vision in great apes is similar to that of humans. They show considerable interest in images, whether they are static or moving. At Jungle Island (Miami), several orangutans in the park have learned to point out various images on a tablet screen when they hear the corresponding words. Linda Jacobs reports that the oldest orangutans at the zoo, Coney (35 years old) and Sinbad (33), showed no interest at all in this type of game. In contrast, Peanut and Pumpkin, twin eight-year-old sisters,[55] were quite fascinated by this activity. Despite the great apes' keen interest in these devices, their enclosures are lacking in technology. Although this is partly for practical and economical reasons, it is also ideological: primates are still seen as wild beings that would risk contamination by our technological objects and our cultures. Indeed, they might risk losing their primary innocence and their "purity" if a watertight separation were not maintained. Many consider that they should be kept in a way that is as close as feasible to their natural state, which, as we have seen, is attempting the impossible. It would be wiser to increase the range of possibilities we offer them—for example, through different activities linked to our technical knowledge.

A Diving Ape

Painting apes, primate mathematicians, and iPad users show that, apart from refreshing latent capacities, new skills are fashioned from diverse opportunities for learning and innovation. Apparently, a chimpanzee learned to dive in this way. A video, shot in 2011, shows Cooper at the edge of a swimming pool.[56] Raised by a scientist, the chimp probably first familiarized himself with the pool, and then learned to move around in the water by progressing bipedally. On one of the videos, the ape can be seen jumping into the pool repeatedly, much like a human child would do. Afterwards, he lets himself sink by throwing his body backwards.[57] He has also learned how to breathe through a diving apparatus connected to an oxygen tank. Before submerging his head, he protects his eyes with his left hand, and dives holding the breathing apparatus in his right hand. Cooper stays submerged, without showing fear at any time, which is astonishing in view of the caution this species usually exhibits towards water. Yet, he seems totally comfortable. Nevertheless, the lack of available information concerning this chimpanzee, the unverified authenticity of the video recordings, and of the scientist taking care of the ape (a certain Renato), call for some caution.

Other Places, Other Possibilities . . .

Living in zoos frees primates from most of the dangers and constraints that they encounter in their natural environment: there are no predators breathing down their necks, and they do not risk being killed by a hunter lying in wait. Food is available in abundance. They have medical care and more time to play or have a nap in the sun, can avoid climatic challenges, have a less complex social life, and are less prone to injury. The enclosures of zoological gardens hence constitute a different world, where great apes no longer have to spend most of their time looking for sources of food and hence have more time for leisure. Furthermore, they are less often on the lookout, as fear of finding themselves in perilous situations is reduced. Additionally, the zoo offers a radically different layout to the natural habitat. For example, in contrast with orangutans in the wild, captive individuals know a *world on the ground*, a world that opens up spaces that are not suitable for use in their natural environment. In the wild, the ground is actually a place fraught with dangers, where only a few vigorous males dare venture. Being able to settle comfortably somewhere with diverse materials within reach without fearing the sudden appearance of an "enemy," and to use their feet, hands,

Figure 3.3 Wattana playing with small balls on the ground, courtesy of Christelle Hano, French National Museum of Natural History, Paris, 2000.

and mouths freely, is something that would be impossible in the Asian forests. In zoos, apes can benefit from these possibilities. They therefore have access to activities that would be incompatible with an arboreal lifestyle. In fact, Jane Goodall herself wonders whether some apes do not lead more comfortable and less stressful lives than in nature, if they live in parks with comfortable enclosures that are designed to fulfill their needs.[58] Besides, most captive primates are now born in zoos. They have thus not experienced what some of their elders had to give up, when they were brutally torn from their world. Nevertheless, it would be a misrepresentation to stress only the positive aspects of living in zoos. The price to pay in exchange for these undoubted advantages is considerable, although it varies greatly from place to place. The moments spent picking and savoring the sweet juicy fruits of a tree in full fruit, a diet with tasty delights varying from season to season,[59] a sunset to be admired in the company of another member of the community,[60] the cozy nest in a wooded landscape, enriched with so many tints of green, unexpected encounters with wandering rodents or reptiles, reunions with conspecifics not seen for a while, the concert composed of a thousand sounds of the forest, the familiar paths, the pleasures of climbing and swinging in trees, being reunited with the other members of the troop in the evening, games among conspecifics, the joy of the hunt,[61] the excitement linked with choosing a sexual partner, the odors, fragrances, colors, the breeze in the air. Everything that makes up an ape's world is lost forever. Moreover, few captive individuals are able to benefit from a wide variety of social contacts, following their own preferences. In addition, when the conditions of captivity are unsatisfactory, one frequently sees stereotypic behaviors, automutilations, refusals to eat, panics, apathy, regurgitation of food, or coprophagy. The risk of disease is increased owing to the stress felt by some individuals, possible contamination by humans, a climate that is not appropriate for certain tropical species, or various dietary deficiencies. All the same, like sanctuaries, zoos represent one of the last places of close cohabitation between humans and great apes, one of the rare places where friendships are formed between representatives of species that are at the same time very close and very different,[62] where intense life experiences can be shared, and where surprising exchanges of skills can occur.

4 An Orangutan Who Can Tie Knots

The animal is first and foremost an organism within its own world, before being an argument in theological debates. What does it do in its world? That is what matters.

—PHILIPPE MULLER, in Von Uexküll, *Mondes animaux et monde humain*, 8

In the 1960s, Heini Hediger, the director of Zürich Zoological Garden, divided animals into two groups: *technophobes,* which avoid culture, and the contrasting *technophiles,* which actively seek it out and try to adopt human technological achievements. Hediger declared that members of the second group "possess faculties of adaptation, and know how to use all manner of technical devices skillfully."[1] When great apes live in close proximity with humans, they have access to a wide range of activities that apparently interest them greatly, and that they therefore try to adopt. In this way, some primates acquire unexpected capacities that were long thought to be unique to humans. For instance, students of ethology are taught that, even though there are many similarities between apes and humans, one of the indisputable differences involves certain fine motor skills such as those involved in tying knots, tapestry-making, or weaving, all activities that exist only within human societies. A single article, written by William McGrew and Linda Marchant, describes a case of knot tying in a group of "wild" chimpanzees.[2] In November 1996, in the Mahale Mountains National Park (Tanzania), they observe a hunt during which a colobus monkey is killed and then eaten by the apes. The next day, they see a few individuals manipulating the remains of the little monkey. A little later, they notice that the female Ako, approximately seventeen years old, is wearing a strand resembling a necklace around her neck. The primatologists retrieve and examine the object, discovering that the leather strip is fastened with a knot. However, this was but a single observation and the knot was seemingly accidental.

The Question of Knot-Making by Great Apes

Robert Mearns Yerkes mentions the topic of knot-making in chimpanzees and orangutans four times in *The Great Apes*,[3] the book he coauthored with his wife, Ada Watterson-Yerkes. The Yerkes present an anonymous account stemming from the circle of Geoffroy Saint-Hilaire, describing an orangutan capable of untying knots.[4] They further note that an individual of the same species is able to *learn* how to tie them.[5] They also cite Furness, who affirms that his chimpanzees do not succeed in knotting the strands that are presented to them, probably due to problems with motor coordination and inadequate perception. Finally, they mention the repeated failures of the female chimpanzee Sallie in all her many attempts to tie knots.[6] Taking these cases together, only one orangutan was capable of *learning* to knot, which would lead the reader to conclude that great apes are *incapable* of tying knots. As Yerkes is considered an authority in the field of primatology, it is a legitimate assumption that his contribution regarding apes' capacity for knotting was influential for some time. More recently, Jacques Vauclair, a leading specialist in animal cognition, claimed that there were no official attestations whatsoever of any primate aptitude to make knots. In his view, this capacity, which involves several levels of difficulty, is lacking from the behavioral repertoire of animals, and nothing supports the notion that a chimpanzee could learn how to tie knots, although a child of two or three can do so.[7]

Nevertheless, several keepers at the Ménagerie of the Jardin des Plantes, and also a number of visitors,[8] had seen Wattana tie several knots one after the other. It was thus not just an accidental occurrence of knotting, but clearly deliberate. Furthermore, Gérard Dousseau thought that Nénette had also known how to tie knots. In order to verify their observations experimentally and demonstrate that some primates are in fact capable of tying knots, I decided, with permission from the director of the Ménagerie, Marie-Claude Bomsel, to offer Wattana various materials that were suitable for knot-making. These experiments were initiated on November 11, 2003, in collaboration with the animal technicians at the ape house. All these trials were filmed.[9] Dousseau recalled that Nénette had made knots with paper, so we first gave Wattana rolls of strong paper. She began to decorate her enclosure with strips of paper, interlacing them from beam to beam. After having played this game for one hour and twenty minutes, she was bored, so she sat down and pushed some paper into a section of bamboo, so that the bamboo was attached to the paper. She then spun it around, by waving the whole contraption, and then, abruptly, *tied a knot,* under the amazed eyes of the visitors present.

Ribbons, Laces, and Other Strands

Over the course of a year, I gave Wattana different types of material[10]: strings, laces, satin ribbons, rubber bands, wool in different colors, laced shoes, pieces of hosepipe, buckskin cords.[11] She made knots with all of them, without being encouraged or rewarded. As soon as materials suitable for knotting were provided, she would start weaving, intertwining, and threading. On several occasions, the opportunity to tie knots was allowed to compete with access to food, yet Wattana almost always preferred knotting. She would usually settle on the floor, and sitting there would develop an unlimited number of variations around a quite particular type of knot: the knot known as an "overhand" or "simple" knot (without the loops of shoelaces).[12] Using this technical know-how as a starting-point, she ties

Figure 4.1 Wattana concentrated on her knot, photo by Chris Herzfeld, Paris, 2003.

single, double, or triple knots, as well as knots of incredible complexity, by slipping cords repeatedly through previously shaped loops. She will also occasionally interlace a string around another stretched out between her two feet. She'll then pass it through again and again, making arabesques, and insert one end of the string into existing knots. She creates assemblages too—for instance, by wrapping a ribbon around a piece of paper, before then closing it with a knot. She also strings pieces of hollow bamboo, wooden beads, or small cardboard tubes along cords or laces. In this way, she twice fashioned objects that resembled necklaces. The first of these, she threw up into the air several times (as she often does with objects she has produced), while she placed the second one around her neck as soon as she finished making it. Wattana also makes use of parts of her enclosure as supports (rings, trunks, or metal gratings). For example, she ties cords to a horizontal beam located at her height, which then allows her to more easily interlace and knot the strands, which—in a way—are then inserted into a "working frame." What she implements then is reminiscent of a type of *weaving*. She is, moreover, capable of lacing and unlacing shoes, even if these are not equipped with eyelets but only very narrow holes, rendering

Figure 4.2 Wattana weaving strings and tying knots on a trunk, photo by Chris Herzfeld, Paris, 2003.

the task particularly difficult. She also links shoes together by knotting their laces. Finally, she sometimes attaches sneakers to a length of tubing, in a symmetrical manner such that each sneaker hangs from one end. She then carries the contraption over her shoulders, much like a water carrier.

Knotting like an Orangutan

One of the most frequent uses of knotting by humans is tying shoelaces. This takes place on a support, the shoe itself, facilitating production of the knots. Wattana, for her part, carries out her activity without the aid of this kind of support. Her "creation" is generally held at the level of her belly or higher up, in front of her face (the strands being attached to some part of her cage). She employs a very complicated system. Being quadrumanous, she knots in an "orangutan manner," using both hands and feet. Her actions are spread across both hands or both feet (left/right lateralization), as well as between hands and feet (high/low). In a way, the whole takes on the role of a *spatial structuring frame*,[13] occasionally including the mouth, which also intervenes in the process as an additional agent. The juvenile ape is even able to tie a knot solely using her lips, as recorded in a film sequence. In a sense, for orangutans the mouth represents a fifth limb, like the trunk for elephants. The manipulations that are thus permitted are so precise that these apes can even maneuver a piece of glass with their tongue without cutting themselves. Wattana thus combines different spatial positions: high and to the left, low and to the left, high and to the right, low and to the right, to which those performed by the lips are added. Although at first sight knots might seem trivial, they in fact involve an extremely complex technique. Finally, it should be stressed that during these sessions the young female is both concentrated and thoughtful. She is able to devote her full attention to this occupation for hours on end. Several researchers argue that this capacity for attention, this perseverance and aptitude for planning ahead, constitute typical traits of orangutans.

On the Correct Use of Knots

When she has finished making an object, Wattana throws it, places it on her head, or undoes it completely. Sometimes she finds functions for her products—for example, in a throwing game or as a "weapon," depending on the circumstances. As a notable example, she tied a shoe to a lace and then twirled it around her head, before hitting her companion, Tubo, with it. She repeated this gesture with a hosepipe she had stiffened with a double knot. In fact, the act of knotting can lend a particular strength and a new configuration to the items that are produced: a piece of tubing thus changes from a linear, soft form to a squat, compact state. Upon completion of her creations, their new properties become visible: the artifact's weight is tested and it is tried out in other ways, perhaps by throwing it up in the air and skillfully catching it a few times, as Wattana would do with a ball. The result achieved is then possibly actively sought during subsequent manipulations. Furthermore, it seems that the materials provided and the objects produced by the young female represent aids for mediation between her and Tubo. These new elements arouse curiosity and give rise to social games. It can,

Figure 4.3 Complicity with the material, photo by Chris Herzfeld, Paris, 2003.

for instance, happen that Wattana produces knots that Tubo systematically spends his time untying, as if they were acting as a duo.

Inventiveness, Rhythm, and Lyric Qualities

Wattana's knots further involve an expressive dimension. Each ape ties knots with a personal style. The gestures linked to knotting seem to constitute a choreography in which rhythm and attributes vary according to the personality of the knot-tying ape concerned. A kind of "connivance," *a complicity with the material*, an *intimate sense of the concrete* (François Jullien, 1992) is expressed; the laces, strings, and ribbons themselves represent expressive materials that participate in this dialogue with the objects of the world. Tim Ingold describes this type of technical activity (weaving, knotting) as being endowed with narrative qualities, in the sense that each gesture develops according to a particular rhythm from preceding steps, and provides the foundation for the next gesture.[14] This delicate encounter between the animal and the opportunities available in its environment leads to creations that display astounding complexity and inventiveness. When she devotes herself to her favorite activity, Wattana never proceeds through trial and error. She links the gestures together without hesitation, apparently following a certain logic, which only she seems able to comprehend. At the center of these physical dynamics, the pace that is imposed on the sequence of movements is the motor that unifies and stabilizes the whole (in which the *spatial structuring frame*, already mentioned, is a stakeholder). Every living being is, de facto, tied to its own rhythm, which guides its ethos, its way of being and behaving in the world. The life of primates is paced by cycles, regularities, a physical rhythm (for example, that of the heart), the alternation of night and day, and the succession of seasons. Although they are often neglected, these rhythms define the tempo of day-to-day existence and are expressed in different forms. In tropical forests, the swaying movements of great apes in the trees are marked by a particular scansion. At Antwerp Zoo, the female gorilla Victoria has acquired the habit of tapping her fingers in a rhythmic manner on the glass panes of her enclosure.

Knotting thus represents an activity favoring the expression of lyric qualities. By contrast, a functional dimension seems to be lacking. The objects from the activity of knot-tying have no functions, do not serve a purpose, and are essentially "gratuitous": *making is sufficient.* For Damien de Callataÿ, "In a fundamental sense, being free of cost implies a connection to the good things in life that is free of painful constraints, that is where the benefits can be gained without any associated pain or trouble."[15] This definition seems to correspond perfectly to Wattana's activity. Despite a perfomance involving impressive complexity, knotting cannot be explained in terms of having practical utility. Yet ethology tends to interpret everything in terms of function and costs versus benefits. It is hence not surprising that this type of behavior, which is unclassifiable and disconcerting, and considered to be marginal and anecdotal to boot, is not included among the research topics addressed by biologists.

How to Learn to Tie a Knot

Numerous scientists and keepers have commented on the fascination that shoelaces have for great apes. In the course of a survey I conducted by email,[16] scientists and keepers bore witness to this. They also explained that apes are capable of *untying* knots. Anne Russon thus described the orangutan's talent for undoing all manner of tangles, no matter how complicated. In this connection, Biruté Galdikas mentions a statement by Louis Leakey: "In fact . . . I have just received a telegram from Dian saying that the mountain gorillas are becoming so habituated to her presence that one is even untying her shoelaces! [Leakey patted his shirt pocket to indicate that the telegram was right there.]"[17]

At that time, the keepers at the Ménagerie would enter Vandu and Wattana's cage to ensure that the apes would grow accustomed to them, and to pamper them and play games with them. The young orangutans lived up to the reputation of apes: they would systematically untie their keepers' shoelaces. The latter were thus frequently forced to *retie* them, and hence to *tie knots,* in the presence of the apes. Wattana would then approach to scrutinize the gestures they would use, with her face just a few inches away from their shoes. Her face was so close that Gérard Dousseau thought that she had problems with her eyesight. In fact, the young female's attitude was simply an accurate reflection of her intense interest in the technical prowess involved in knotting. As noted previously, the actions of carefully studying an activity, and then trying to repeat it, represent a privileged form of learning, common to all higher primates (from which comes the famous "Monkey see, monkey do"). After this observation phase, Wattana would attempt to replicate the sequence of gestures that she had carefully observed, repeating them and linking them together, but without managing to tie knots. She continued to practice, seeking to observe the movements of her keepers as often as possible, before trying again to repeat the moves as soon as she could get hold of a piece of string or rope. In this way, she progressively assimilated the process, progressing by making a series of adjustments and readjustments, until the moment came when *she succeeded in tying her first knot*. Danielle Dousseau, professional keeper at the Ménagerie and the daughter of the former head keeper, recalls seeing Wattana tying a knot when she was only four years old.[18] She thus *incorporated* the movements of knotting rhythmically, perfecting her know-how little by little and refining her gestures, in just the same way as a human child learning to tie knots by imitating its parents.

When she devotes time to this activity, Wattana does not seem to have a precise, predetermined, or schematic idea of the motions that she wishes to accomplish. She instead uses her visual memory—thus transposed from one sense (vision) to another (touch)—as well as *body memory*, something that pianists, for instance, know well. Although musicians initially decipher a score note-by-note, they break away from this little by little, subsequently using it only as a memory aid. At this point, their fingers take over. Thanks to multiple repetitions, each position of the fingers on the keyboard commands the next. Memorization occurs at the level of the hand, with rhythm in support. The process described here does not match the Aristotelian principles of *production theory*, which represents the body as a subordinate of the mind: the artisan manufactures a chair by *mentally* following the production plan for that object. After all, one does not learn how to tie a knot by obeying instructions for use dictated by the intellect that determine the movements to be executed: knotting is a manipulation that is far too complex to fit such a scheme. Knotting notably requires integration of gestures that must be linked together in a subtle rhythm.

Of the Art of Imitation

Examples of apes imitating humans have been reported since Antiquity: primates take on the function of a servant, dance, or eat at the table with their host. In the eighteenth century: "At Reims, a man presented an 'Academy' of dogs and monkeys costumed as soldiers and labourers that attacked a town, played cards, danced and ate at a table. Monkeys enjoyed pride of place. At

Paris, in 1714, one monkey—dressed as Harlequin, a musketeer or a young lady—saluted the company, sat on a chair and performed infantry exercises and acrobatics. Another parodies the convalescence of a famous actor and achieves great success, in 1767."[19] Wattana herself is an excellent imitator. One of her keepers, Émilie Gouverneur, told me how she had shown her how to weave herself a "bracelet" using the green or yellow plastic nets used, for instance, to package lemons. When Émilie returned to the ape house a little later, she found Wattana with several nets around her wrists and forearms. One morning, I rewound the wool that I was planning on giving her into a small ball in front of her. Wattana was fascinated by this rewinding action. During the following experimental trial, she suddenly started performing the same gesture, the first time with a very slight trace of hesitation, but then continuing the movement in an assured manner. According to her animal technicians, she had never practiced this "technique" previously. Yet she very easily imitated this series of gestures, although they are quite complex because they require precise coordination of movement. A young child would have trouble achieving as much at the first try. This example shows that Wattana is capable of reproducing a complicated gesture with great ease, without having to go through a procedure of "trial and error."

In his response to our study, the renowned primatologist Andrew Whiten writes that "knotting" activity represents an interesting theme for exploring the topic of imitation.[20] He actually had the opportunity to observe chimpanzees attempting to tie knots and considers that they proceed by imitation. Orangutans described as being in "rehabilitation"—that is, in the process of being reintroduced to their natural environment after a period of captivity—bear witness to this eagerness to imitate certain human behaviors and acquire new abilities.[21] Anne Russon provides various illustrations of this tendency. In her center, most apes know how to use cutlery. Davida stole a toothbrush and toothpaste, brushed her teeth, and then cleaned the brush in the same way as one of the volunteers at the camp. Princess imitates the women that wash the laundry in the river. The primatologist also mentions the case of the young orangutan Pegi, who was ill and at first refused to eat the grape that was put in her mouth, but then accepted it after seeing Russon eat some of the grapes. Behaving cautiously, she did not want to eat a food item that was unfamiliar to her without having been shown that it was safe by a more experienced individual.[22] Finally, she discusses Supinah, who was more inclined to acquire certain skills from humans than from her conspecifics. This young female had actually spent her whole childhood with Biruté Galdikas, who reared her like a human child. Supinah knows how to drive in nails or use an axe, she saws wood, sharpens blades, digs with a spade, sweeps, paints buildings, and pumps water. After having seen the cook light the fire with kerosene every day, she managed to do the same, and kept the fire ablaze by performing a fanning action over the embers with the lid of a dustbin. Later on, she unhesitatingly grabbed a burning lump of wood in the hearth before trying to use it to set fire to a cup of fuel. These behaviors are all the more impressive since the capacity to manipulate fire has long been thought to be something that truly belonged exclusively to humankind. Numerous other examples of imitation have been observed by Russon at Camp Leakey[23]: the orangutans drink tea or coffee; use spoons, cups, or plates; attempt to prepare or cook food; wash laundry; siphon off gas from boats; and write on memo pads. Biruté Galdikas[24] has also described this kind of appropriation. For instance, she recounts how one of the camp's orangutans broke some eggs, put them in a bowl with flour, and then mixed everything. Some apes even try to start the engines of the boats. Talking great apes are, of course, not to be outdone. For example, Lyn Miles, the instructor of the orangutan Chantek, recalls the ape playing "Simon says,"[25] and easily reproducing various expressions, tapping with his feet, jumping, raising his hands, blinking, and banging his chest. Miles demonstrates that these behaviors fulfill the various criteria for being viewed as "true imitation." She underlines the importance of the cultural context, which is fundamental for imitation. Apes that were raised in human families thus benefit from marked *social emulation*, which plays a key role in the learning process.

Passing on Knowledge

Rendered more dynamic by this social emulation, the privileged mode of learning of apes involves observation (higher primates need to *see*, and even *watch carefully* in order to learn),[26] imitation and reinterpretation of the behavior they wish to acquire. In captivity, keepers serve as role models, *charismatic leaders*,[27] taking on major importance for their charges, who are fascinated by their world. Furthermore, mirror neurons fulfill an essential role in this process. When Wattana watches her human companions tie knots, specific neurons are activated, the same as those activated in the individual she is observing carrying out the action to be replicated. In a way, despite never having carried out the actions she scrutinizes, she thus already possesses a form of "cerebral experience" of the movements that she aims to adopt, as the neurons involved are stimulated as soon as she tries to repeat the same series of actions. Although it plays such a major role in the transfer of knowledge and skills, imitation is disparaged. It is still thought of as something servile and stupid, with the imitator being considered a deceiver. Indeed, the phrase "to ape" has very negative connotations. Yet, when it comes to learning how to tie knots, it is not just about reproducing gestures in a mechanical manner. Rather, learning requires commitment to a subtle dynamic, which includes tactile, cognitive, social and cultural dimensions, and the operation of a complex mechanism, that of the mirror neurons, as well as the aptitude to transfer the information collected from one mode (vision) to another (touch). Beyond simple imitation, there is an interpenetration of worlds, a transfer of ethos, and a reappropriation of knowledge and technical abilities.

Practice, Testing, and Incorporation

Knot after knot, assembly after assembly, weaving after weaving, Wattana repeats the chain of movements involved in knotting. Little by little, she refines her gestures and increases their complexity, developing true technical mastery. For the young orangutan, each ring and post in the enclosure becomes a support for knotting, a weaving loom. This intense practice, during which she tests, improvises, and gradually incorporates a modus operandi in a creative fashion, is simultaneously a rhythmic and a hand-to-hand struggle with the world and its materials, colors, consistencies, achieved with the aid of fibers and cords. Precise and completely dedicated to her task, Wattana knows exactly what she wants to do and how to achieve it. She grabs everything that can be used to tie knots, including materials that seem unlikely to be suitable. Sometimes she is absorbed in this activity for a whole afternoon, and thus expresses a real *taste* for the execution of knots. Her sustained attention, the depth of her involvement, and her *craving to do it right* all testify to this.

Pleasure as a Motor for Expertise

As her experimentation progresses, Wattana's knotting abilities get better and better, and she seems to derive more and more pleasure from knotting. Stimulated by recurrent provision of new materials and varied accessories, her attraction for knotting increases at the same time as her proficiency develops further. This is when a "pleasure of doing," a Funktionslust,[28] emerges, something that is defined by the Belgian phenomenologist Marc Richir as the pleasure felt by some living creatures when they *do what they know how to do*, when they *do what they know how to do well.*[29] Funktionslust is also the "functional enjoyment" animals derive from exercising their physical faculties: the pigeon taking flight when it stops raining, gibbons swinging elegantly from tree to tree, the cat crawling through the tall grass on the lookout for prey, the dog attempting to catch a frisbee, the dolphin gliding harmoniously through the ocean waters. This functional enjoyment is also fed by the *pleasure of doing what one knows how to do well.*[30] This feeling is also described as "bodily enthusiasm,"[31] "inebriation,"[32] "spontaneous, picturesque and free virtuosity,"[33] or as " extravagant motor activity," in the words of Étienne Souriau. Thus Funktionslust also plays a role in accelerating learning. The pleasure that is felt increases in magnitude when carrying out an activity that one is able to perform with increasing aptitude. This encourages the individual to pursue this activity, while reinforcing skill. Expertise (and in some case a true virtuosity) develops, carried by the mounting jubilation felt by the individual who is engaged in this process. The urge to knot is so intense in some cases that great apes find ways of obtaining the materials needed for this activity. For example, Wattana tore up a tee shirt that she had received from her keepers into long strips. Using the strips that she had thus fashioned, she then made knots. At Lowry Park Zoo in Tampa (Florida), the female orangutan Deedee used palm leaves that were present in her enclosure, which she used to tie knots after first cutting them into strands. She also used bark and grass to fashion various artifacts. When visiting the zoo in Atlanta, Lyn Miles employed sign language to encourage Chantek to tie a knot. He immediately searched for something suitable, and grabbed a blade of tall grass from his outside enclosure. At Houston Zoo, Bubba made knots with the hair on his forearms (these knotted hairs, called *dreadlocks*, can be very long). Some apes divide the ropes in their cage into several strands so as to have access to materials for knotting.

What Is It like Being an Orangutan When You Live in a Zoo?

Towards the end of the 1970s, the primatologist Jürgen Lethmate[34] performed thorough studies of the behavior of captive great apes. He discusses the common tendency to think that chimpanzees are superior to the other species, notably when it comes to their capacities for manipulating objects. He further underlines the excellent propensity and astounding performances of orangutans in this regard. Lethmate notes that they twist, knot, wring, and attach ropes, wool and wood, clothes, pieces of cloth and gunny,[35] so as to create constructions on which they can swing, just like they do on a hammock.[36] While evoking this penchant for manipulating and assembling different components, he points out that it is more pronounced than in chimpanzees.[37] Additionally, orangutans appear to be able to use accessory tools, such as a rope, or a stick, to reach other objects. In an article published in 1982,[38] Lethmate explains that the orangutans display outstanding ability for using a variety of tools. He highlights their aptitude for utilizing, and also for building, practically all the tools that had hitherto been described as being used within chimpanzee communities. Finally, he notes that their fascination for sophisticated manipulations (their predilection for shoelaces being one example)[39] is far more pronounced than in other primates. This also applies to their greater aptitude for using one tool to create another. As for the keepers I questioned, they explained that orangutans are deliberate, meticulous, patient, forethoughtful, persevering, and endowed with great dexterity. Like the researchers, they confirm that orangutans are capable of intense concentration for long periods, and can plan actions ahead over several days in order to achieve a given goal. They also exhibit a marked taste for solitary activities. These tendencies are tightly linked with millions of years of coevolution between these Asian great apes and the humid tropical forests of Borneo and Sumatra. As they only get very few opportunities to exert these capacities in captivity, the apes attempt to do so as soon as they have the possibility in human-ape communities. Yet, the fact that they live with humans entails encounters with knots. These thus become an opportunity, which allows them to exercise their forest skills.

From Forest Skills in situ to Knotting Skills ex situ

Plaiting, interlacing, intertwining: these are all terms used by primatologists to describe construction techniques that great apes use in building nests. Branches, leaves, and twigs are *woven* into a confortable refuge for the night, or for a nap. During the e-mail investigation that I conducted at the beginning of this study, scientists immediately linked the orangutans' sophisticated nest-building techniques with their capacity for knotting: when they make their beds, they employ the same gestures, the same capacities, and the same materials (i.e., fibers) that they use when they tie knots. Several primatologists additionally noted that, in their natural habitat, they also show a certain propensity for attaching, joining, and assembling separate elements. Moreover, these tendencies are stronger than in their African cousins, as the arboreal lifestyle is more pronounced in the Asian great apes. Great apes are not satisfied with building a basic bed. They decorate their nests with a "plant mattress" that some researchers consider *artistic*. They select their materials according to the available plant life and their shades of green. They then line their nests with *Campnosperma* branches to protect themselves from mosquitos. This plant is also used for the same purpose by the Dayak people, who live in Indonesia and in Malaysia, in the same forests as the orangutans. Some apes systematically bite the tips of the branches used for the fringes of their nest, the edge being composed of twigs of similar appearance and of the same length. Some of them fashion a small cushion from plants that they grip tightly against themselves when they sleep. Apes also seem to attach importance to the panorama that can be seen from the bed: they choose the site carefully according to the view. Their shelters may also be equipped with "roofs" to protect them from sun or rain. Sometimes, this kind of construction can serve as a structure permitting passage from one point to another like a bridge (*bridging nest*)—for instance, between two trees located on opposing banks of a river.

According to Carel van Schaik, orangutans understand the basic principles of nest construction, structures that allow them to have some degree of control over their environment. In a sense, the knots that are made in captivity bear witness to the complexity of the nests constructed in the wild. The virtuosity, the complexity, the inventiveness, and the spontaneous dimension of the knots fashioned by Wattana help to draw attention to the sophistication of these nests, to their technical qualities, and to their undeniably aesthetic dimension, leading us to view them with greater respect and admiration. Indeed, they cannot simply be reduced to their function alone. In fact, it is surprising that what the leading American specialist of apes, Robert M. Yerkes, describes as *constructions* have not been studied more intensely in terms of their existential importance for all apes.[40] The term "nest," reserved for the shelters birds make with their beaks (which are by no means less sophisticated), does not differentiate constructions that are built by hand. Yet, the beds fashioned by great apes seem closer to certain *constructions* that are erected in elevated locations by certain peoples. For example, the *tree houses* of the Korowai of Papua consist entirely of plant materials and are built in the treetops, about ten meters above ground level. The day nests, which are constructed on the ground by the three other great ape species,[41] might be seen as resembling huts with a triangular base. Although they differ in many respects, these constructions represent systematic attempts to adapt and modify components of the environment in a harmonious manner, so as to *create an existence* in a given habitat. They are lightweight architectural structures, which are comfortable and non-polluting, do not destroy the habitat, and are very quick to produce.[42] These artifacts are crafted with fibroconstructive techniques, that is to say, with methods that are highly complex as regards technical and structural aspects, and with fibers as the material of choice.

Remarkable Practitioners of the Forest

All apes have experienced a long and slow coevolution with their forest habitat, where *fibers* are ubiquitous: trees, creepers, bark, leaves, fruits, berries, mosses, branches. For them, the nurturing forests are plant kingdoms that provide of all the necessary components to *make a life for themselves* and satisfy their various needs: rest, body care, comfort, sex, diet, medication, games, protection, and attack. The natural components that they use are predominantly made of fibers and allow the construction of nests, various devices, comfort aids, or tools. In line with local traditions, marked by the techniques, skills, and objects of the orangutans living in the same region, the apes fashion plant cushions, protective gloves made from leaves, leafy towels to clean their faces, napkins to wipe their chins, roofs to protect against the rain, and screens to shield them from the sun. They use leafy branches to chase away flies or wasps; combs made of plant material; fans made of leaves to cool themselves; leaves for cleaning their bodies, to pick off lice or to squash external parasites; plant sponges; walking sticks, or even sticks to pick their noses. In this way, they invent commodities that are so beneficial that they become habits that are passed on, and then traditions. Consulting the first two coauthored articles devoted to the material cultures of chimpanzees and orangutans[43] reveals that the majority of behaviors, devices, or tools (associated with body care, sex, games, comfort) belong to what the French philosopher Dominique Lestel would describe as an *ethology of comfort*. In a passage devoted to Hans Jonas, he writes: "Interiorization originates with the beginnings of life. Each organism reaches awareness for itself, through the fundamental interest that it feels for protecting its existence and prolonging it. Its life is selective and 'informed'. It is never purely reactive."[44] Further on, he adds that "the means that the organism mobilizes for its survival (sensory perceptions, feelings, intelligence, will, the power to command its extremities, the choice of goals) should never be considered as means serving an end, but always as qualities of the life to be preserved."[45] The *concern for self by self* and the *simple fact of existing* constitute tasks that our societies tend to forget, overshadowed by the sophisticated projects that we choose to implement. Great apes experience a *power* and a *pure pleasure of being*,[46] which carries them beyond mere basic needs; to advance towards what for them represent possible sources of comfort, satisfaction, or pleasure.

Great Apes as Fiber-Users

How can one explain the lack of interest shown by the scientific world in the fibroconstructive techniques and items of comfort that are omnipresent in the world of great apes? According to the first articles dealing with the cultures of chimpanzees and of orangutans, elements fashioned from fibers account for about 92% of the objects, devices, and tools that one finds in their world. The difficulty of studying objects manufactured using fibroconstructive techniques (it being necessary to climb up quite high to examine the nests, and objects made out of plant materials are ephemeral) is not sufficient to explain this indifference. "Tool ideology" (Kenneth P. Oakley, 1964) undoubtedly played a role in overshadowing all other aspects. In the article "Axes of Perfection: Stone Implements and the Predicament of Progress in 19th Century Prehistoric Archaeology,"[47] Nathan Schlanger demonstrates the extent to which tools constituted the cornerstone of a nascent discipline in the nineteenth century: the study of prehistory. Providing reliable evidence of hominization, they are perfect mediators between the very distant past and the present.[48] Moreover, the direction taken by the discipline is built on the underlying sweeping dichotomous categories of male versus female, of the hard versus the soft, of the public versus the private (intimate) domains. Fibers are considered to be *soft* materials, in contrast with stone, representing the *hard*. A great divide exists between, on the one hand, the intimate and comfort (traditionally placed on the feminine side), and, on the other hand, the technical and domination of nature. All this seems to have implications for the question of culture in the great apes, as well as for numerous other sectors of primatology. Here, too, the history of the discipline seems to have been marked by the omnipresence of the tool. Yet material cultures of both ancient hominids and great apes are by no means limited to the stone tools beloved of prehistorians: they constitute only a minute fraction. A major part of the cultures born within the forest environment is naturally linked to fibers and the use of fibroconstructive techniques. Paying greater attention to knots and nests thus adjusts our thinking towards a paradigm where the use of fibers is taken into account, and where the human is an *artisan* and a *builder*, rather than a *maker* and *tool user.* As proposed by Nold Egenter, it would surely be very fruitful to ponder the question of hominization using *construction* as a starting point. But the questions and options prescribed by prehistorians are so dominant that methodological tools for exploring these dimensions are lacking. Primatologists align themselves with their prescriptions and give precedence to use of a stone tool to break nuts on a natural anvil, which is admittedly impressive in its proximity to the stone tool industries of ancient hominids. Nonetheless, primates are clearly "fiber users," rather than "tool users." For Tim Ingold, activities linked to weaving are antecedent to those concerned with *production*. From his viewpoint, "*making*" is a modality of "*weaving*": "Only if we are capable of weaving, only then can we make." The skills required for constructing nests and for working with fibers would thus have allowed the development of the physical and cognitive capacities that are indispensable for the production and use of tools. Activities connected to crafting with fibrous materials also seem to have kick-started a diversification of behaviors, a particular care devoted to the choice of sites and of materials, the development of transport strategies for the necessary components, and assembly techniques. Many captive great apes manipulate different types of fiber: palm leaves placed on the back, plants positioned on the head, leaves used to squash small insects, branches inserted into artificial termite nests, fruits used as containers. For instance, Jordan describes the case of a bonobo that used half a red pepper to scoop up some water.[49] At the zoological garden in Planckendael, a female of the same species, Dzeeta, devised the idea of using a pineapple (which she had cut into two pieces herself), as a receptacle.[50] Similarly, Nénette tore a fragment of wood from a trunk in her cage, and used it to reach some grapes placed high up on a mesh grating by the keepers. She fashioned her tool to reach the fruits by moving them from a part of the mesh that was too fine for her fingers to pass through, along to a section with wider mesh.

Knots: Nueva on Tenerife and Meshie in New York

Different forms of manipulation of fibers are thus found in captive great apes. Some past accounts already demonstrated the enthusiasm with which great apes greet knotting. Wolfgang Köhler, best known for his research on the capacity for tool use of chimpanzees (conducted on Tenerife between 1913 and 1920), recounts how one of the chimpanzees at his center, the female Nueva,[51] was perfectly capable of tying knots. She devoted a great deal of time to collecting and assembling various objects. She tied a wool rag around a stick. She intertwined and plaited pieces of straw into the slits of metal grilles. She pushed a banana leaf through the cagewire, recovering one of the ends (with some difficulty), and pushing it through another opening so as to tie both ends together. She repeated the procedure, either by tying a second knot, or by passing the leaf through the already tied knot. In Köhler's opinion, she made a deliberate effort (even if rudimentary) at *construction*, performing a kind of *manual labor* akin to that performed by a craftsperson. Where Nueva was limited was in the impossibility of persisting with her projects and of planning over the long term. These limits play a lesser role in orangutans, which show greater aptitude at organizing their actions over the medium or even long term. Nonetheless, Nueva showed real passion for making knots, and a dedication to this type of activity, a deep interest and a *joy of making*, isolating herself from her companions for long periods.

In another example, Harry C. Raven, curator at the Department of Comparative Anatomy of the American Museum of Natural History in New York, recounts in the Museum's magazine the adventures of the female chimpanzee Meshie, who was four-and-a-half at the time.[52] Meshie had lived for a year in Raven's company in Cameroon, where he had procured her in 1930 before subsequently bringing her back to the United States. The young chimpanzee provides an excellent illustration of the great ease with which great apes acquire various human skills. Among other things, Raven describes her talent for freeing herself from any predicament. Meshie first found out how to undo knots by pulling on the cords with her fingers, then eventually with her mouth, and thus became an expert in the art. He also reports that she occasionally managed to *make knots,* but in a rough-and-ready manner, much like children when they are learning to tie their shoelaces. Raven adds that the most surprising feature was the obvious pleasure she derived from this occupation.

The cases described are, however, relatively rare and no systematic studies have been conducted on knot-making by great apes. Nonetheless, my own study (96 meaningful responses) does strongly suggest that various great apes are capable of tying knots, whether they are captive, trained to "talk," reintroduced, or integrated into human families. Even if she is exceptionally talented, Wattana is far from being unique.

Knot-Tying Apes

During a study carried out at the Ménagerie of the Jardin des Plantes, Gérard Dousseau told me that Nénette also showed the ability to tie knots. Incidentally, during the experiments performed with Wattana, Nénette also made several knots, in her own way, slowly and using only her hands.[53] At Taronga zoo, in Sydney, the orangutan Judy displayed brilliance in terms of both her aptitude for knotting and her ability to untie knots. She showed incredible dexterity and used whatever she was able to find: ropes, jute bags, banana leaves, or other fibers, such as bamboos or long grasses. She often tied her knots onto the cagewire and also loved to paint. Lisa H. Abra, animal technician at the Taronga Zoo, was able to enter Judy's cage without problem, as the ape was exceptionally mild-mannered and calm. She was born on October 26, 1957, and was raised by her own mother, but only until the age of two. Humans then took over. She died on June 10, 2007.[54] Terri Hunnicutt, keeper of primates at Saint Louis Zoo, told me that Junior (captured in 1962 and deceased in 2006) was excellent at tying knots. He would use ropes, hosepipes, or pieces of cloth and would make both simple and more elaborate knots, some of which resembled long braided cords.

Junior would tie knots with his hands and with his mouth, and seemed to greatly enjoy this activity: as soon as he caught sight of a rope, he would approach. He was hand-reared and was very well liked by the keepers at the various zoos where he spent time (Tulsa, Houston, Memphis, Saint Louis). The female bonobo Dzeeta at the zoo in Planckendael (born in 1971 and deceased in 2002) also knew how to tie knots. After her birth in 1971 in the Democratic Republic of the Congo, she grew up in a human family until the age of six. She had adopted the habits of her companions and had learned to knot.[55] As mentioned previously, Bubba, a Bornean orangutan, had found a way to tie knots using a commodity that he constantly had "at hand": the long hair on his forearms. This he would, by necessity, knot with his mouth and lips. Afterwards, he would present these bodily embellishments to his keepers by pressing his arms up against the fencing. He thus succeeded in arousing their interest and in engaging them in social interactions. Nevertheless, Lynn Killam, supervisor at the zoo in Houston, was only able to observe this behavior on two occasions.[56] Bubba died on April 23, 1998. At Lowry Park Zoo in Tampa (Florida), Deedee[57] also devoted herself to the activity of tying knots. According to her keeper, Lorin Milk, she would use both her hands and her feet for knotting. In contrast with Wattana,[58] she apparently did not use her mouth. Milk also reported that Deedee tied a small sheaf of long grass with three strips of blue cloth (distributed in perfectly symmetrical manner), which she had previously torn from a T-shirt. Milk kept this combination, along with various other products, which she took care to photograph. Deedee's creations are impressive in their simplicity and harmony. I was later fortunate enough to directly observe her knotting.[59] After having prepared her own knotting materials by tearing strands from palm leaves, Deedee crossed the ends of these "leaf strips," and then slipped one of the two into the loop she had shaped, while holding it with her feet. She then pulled the strands with both hands to close the knot. She repeated these gestures several times, and then later encircled some palm bark using a strand made of plant material, and closed this with a knot. She thus fashioned an *artifact* that her son Berani grabbed as soon as she placed it on the ground. In Fresno, California, Siabu (born on March 25, 1989), also an orangutan, ties knots several times a week. She was raised by hand,

and lives at the Chaffee Zoo.[60] When I visited her, she immediately tied a knot into the ribbon that I gave her. In addition, she very kindly took care of the baby of the female siamang that shared her enclosure. At the Metrozoo in Miami, several orangutans are able to tie knots: notably Jasper (born in 1973), Butch (1985), and Monja (approximately in 1952). The latter lived until the age of 55, representing a longevity record for the species. It would seem that Satu (born on August 31, 1999), a female orangutan at Dvur Kralove Zoo (Czech Republic), is also proficient at knotting,[61] and the same applies to Mei (1985) at Monkey Jungle in Miami.

Moja, Chantek, Panzee, Panbanisha, Kanzi, and Princess

Turning to *talking apes,* I contacted the best-known researchers in this field, to find out whether these primates were also adepts of knotting. Some of them were indeed also able to tie knots. Among the group of chimpanzees kept by Roger and Deborah Fouts (Human Communication Institute, Ellensburg)—which include the famous Washoe, the first great ape to have had an exchange with humans using sign language—this aptitude was observed only in Moja.[62] She would tie shoelaces, but only in a loose manner.[63] She was also known to have produced several figurative paintings. According to Lyn Miles, Chantek (whom she raised and trained to express himself in sign language) would engage in what the psychologist describes as *Arts and Crafts*. He would knot cords and string them with beads, wooden letters, or other objects, creating hanging ornaments and various macramé constructions. He would also fashion necklaces with the striking feature of being adjustable. Miles thinks that he had noticed that she wore necklaces of different lengths and that the human beings she encountered had necks of different sizes. On his own initiative, he had developed a system of sliding knots. This type of construction constituted about a quarter of his output.[64] When Lyn Miles and I went to visit him at the zoo in Atlanta, the primatologist asked him, using sign language, to make a knot. The large male was sitting in an enclosed grassy area. He nevertheless found a way of complying: taking hold of a long grass stem, he fashioned a tiny knot with his large hand. At this point, Miles threw him a plastic bag with some beads and some string. He then strung a bead on the cord before tying a knot.[65] According to the journalist Eugene Linden, he is also capable of producing a knot using only his mouth.[66] Born on December 31, 1985, the female chimpanzee Panzee (also called Panpanzee) was raised with the bonobo Panbanisha at the Language Research Center (L.R.C., Atlanta), allowing researchers to compare the learning of iconic language in the two species.[67] Panzee tied a rope to the wire netting of her cage when I visited her at the L.R.C.[68] She also acquired other human skills. For example, she would trace small signs (resembling shorthand symbols) in a notebook, while "writing" very carefully on the lines without deviation. In the same center, the bonobos Kanzi (born on October 28, 1980) and Panbanisha (born on November 17, 1985) are also able to tie and untie knots.[69] According to Francine Patterson, the gorilla Koko (born in 1971) *tries*, from time to time, to tie knots. She attempted to tie Patterson's shoelaces after seeing a video ridiculing this procedure.[70] However, Patterson does not clearly affirm that Koko is able to tie knots. Finally, in the sanctuaries where attempts are made to rehabilitate great apes for life in their natural habitat, some apes also tie knots. For example, Princess, of the Tanjung Putting National Park (Indonesia), learned sign language with Gary Shapiro. Anne Russon affirms that Princess knows how to tie knots.

All knot-tying apes[71] are hence highly encultured, whether they are rehabilitant apes, talking apes, or apes living in zoos. So all of them were reared by human beings. Of the sixteen cases of knot-tying apes that I was able to discover[72] (twelve females and four males), eight were orangutans; three were bonobos; and five chimpanzees. Most of these apes are therefore orangutans (50%) and females (75%) living in zoos. The zoo environment offers them new opportunities, providing access to activities that interest them enormously, and to which they can dedicate sufficient time.

What Is an Orangutan?

We must remember that orangutan enclosures are generally devoid of trees or any other element typical of the forest. The apes thus have no possibility to implement their talent for mechanical engineering, their remarkable mastery of tools, or their capacity for resolving complex problems.[73] As noted previously, orangutans have developed sophisticated techniques for coping with living in forests because of the constraints imposed by their environment (where resources are rare and difficult to reach). When they reside in zoos, these Asian primates are hence particularly interested in the opportunities humans have to offer them: those that allow them to practice their skills, to express their propensity for manipulating objects, and to exercise their cognitive abilities. Gestures related to the fibroconstructive technologies of "wild" orangutan populations are in this way employed in captivity, but in a different mode. Moreover, great apes are genuinely fascinated by techniques used by the humans that surround them, as well as by their most advanced technologies, as demonstrated by the apes that use iPads, touch screens, joystick games, or computers. Cohabitation of humans and great apes further allows the enhancement or emergence of particular skills, and the expression of an inventiveness that was long ruled out for primates, thought to be mediocre imitators of mankind. Wattana's knots testify to this creativity, which is also present in animals that are less closely related to us. Thus, the possibilities generated by the biological are expressed when conditions are present that favor their emergence. We are "brought back over to the totality of the organism" (with its structural specificities and its constraints, its limits and its potential, its life story, its way of being in the world, of perceiving, of exercising its cognitive abilities, as well as the aptitudes of its species) ". . . and to the entirety of its relationships with its concrete or experienced habitat."[74]

The Knowledge of Keepers

In the context of this study, the expertise of the zookeepers has been essential. Indeed, they were the first witnesses of knot-making by Wattana, Nénette, and Deedee, as well as of many other apes. These keepers, these primate specialists, have provided the cases mentioned, the examples given, along with various reflections and insights. These genuine devotees possess impressive expertise. Moreover, they defend a unique and original point of view. Furthermore, one cannot contemplate carrying out observations or experiments in zoological gardens without their precious aid. Their relationship with the animals they care for is enriched by close proximity developed through daily contacts and shared slices of life. Despite this, however, the profound knowledge of animal technicians is often ignored, or even discredited. Scientists discount keepers, precisely because of their close links with the great apes, as these links violate the requirement to maintain distance between research and the object of study. In fact, keepers fulfill a key function in the apes' lives, playing a parental role (feeding, cleaning, care connected to the intimate sphere) that sometimes goes as far as bottle-feeding their babies. As a result, they are in close contact with them through vision, sound, smell, and touch. Jocelyne Porcher, who has done much work on the links between caretakers and "their" animals, argues that this proximity also, and perhaps *especially*, occurs through the *body*.[75] Accordingly, the personal experiences of keepers are also of a physical, corporeal nature. Their relationship with the apes is woven out of trust and respect, and their great understanding, their attentive involvement, combine to ensure the originality of their knowledge: "The study of animal behaviour demands from the observer such close intimacy with the living animal and such superhuman patience that a theoretical interest is not enough to maintain the necessary level of attention, without the love that correctly reveals the kinship in the behaviour of man and animal that was intuitively felt."[76]

On the Aesthetic Sense in Great Apes

5

The brown stagemaker (*Scenopoeetes dentirostris*) lays down landmarks each morning by dropping leaves it picks from its tree, and then turning them upside down so the paler underside stands out against the dirt: inversion produces a matter of expression.
—GILLES DELEUZE AND FÉLIX GUATTARI, *A Thousand Plateaus*, 315

Wattana's Knots

What we would describe as an "aesthetic sense" is expressed during the knotting activities performed by apes. Laces, strings, and ribbons given to Wattana then become expressive materials. The choreography of knotting itself represents a form of expression. The complicity that develops with the various materials, a sensitive encounter between the animal and the opportunities provided by its habitat, leads into eloquent shapes of an unexpected complexity. The rhythms adopted by each individual ape become a working environment which harmonizes the whole and determines a first perimeter, a world, a territory, providing a certain degree of autonomy. When great apes tie knots or fashion their nests, their gestures are those of the artisan: twisting, bending, lacing, rolling up, weaving, folding, knotting, attaching, crossing, intertwining, braiding, wringing, interlacing, and joining.[1] In a way, the great apes do in fact produce *artifacts,* in agreement with the definition of the term,[2] and even objects representing a form of arts and crafts,[3] according to Lyn Miles. The knotted objects and the constructions fashioned by the great apes are thus productions that are linked to the know-how and skillfulness (which mature into expertise) of the individual who fashions them, a *form* of handicraft. Wolfgang Köhler goes further still by describing these activities as "artistic manual labor." The ape practitioner displays care, judgment, dexterity, and clearly shows a concern to do well. Through playful exploration of the possibilities afforded by a technique that is mastered to perfection, creativity is expressed and preferences revealed for colors, materials, and shapes. Letting their inventiveness come to the fore, great apes can transform one of their favorite materials, *fibers* (whether these are strands that can be used for knotting or various plants for making nests), into objects that testify to their sense of harmony and their attraction to what we would describe as "beauty." Moreover, the determination, the concentration, and the perseverance exhibited during these intense moments of "creation" serve as a perfect incarnation of the notion of Funktionslust. The knot-tying apes truly seem to derive considerable pleasure from engaging in this kind of activity. Indeed, sometimes a kind of restlessness becomes apparent, as if there were some *urgency to take action.*

Among the knots fashioned by Wattana, two are worthy of particular attention. Firstly, the young female draped long strips torn from a roll of white paper everywhere in her cage, attaching them to several of the beams high up in the enclosure, as if the paper were streaming from top to bottom. At the end of this exercise, she had in a way created an *installation* through which she joyously passed to and fro. Secondly, on a different day, she suspended more than a dozen very long sections of wool of different colors from a strut in her enclosure. She then knotted the strands, intertwined and interlaced them, weaving them in a way. The magnificent blending of blues, purples, mauves, the shapes generated, the harmony of the whole, in my opinion made it an opus. Linked symmetrically by three blue strips,

Figure 5.1 Wattana's knots, photo by Chris Herzfeld, Paris, 2003.

the bundle of long grass fashioned by Deedee is also worthy of mention, for its simplicity and its precision. Distinguished through marked differences in style, frequency, knot type, rhythm, use of foundation, and methods of knotting, the artifacts of knot-tying apes further portray a particular way of existing in the world.

Great Apes Like to "Create Art"

The aptitude of primates to take over certain elements of human know-how and to manifest aesthetic sense is not limited to knots. Primatologist, artist, and author of the best-seller *The Naked Ape*,[4] Desmond Morris enumerates around thirty cases of primates engaging in painting in the 1950s, the "Golden Age of ape painting," in the United States, in England, in the Netherlands, and in the German-speaking countries. In *The Biology of Art*,[5] he analyzed the works of twenty-three chimpanzees, two gorillas, three orangutans, and four capuchin monkeys. During the initial sessions, the primates were initiated in the use of a pencil or paintbrush. Afterward, they were left to their own devices to experiment: "During a short apprenticeship Morris placed a pencil in the chimpanzee's hand and pushed the pencil over the paper, holding the monkey's hand in his. Once Congo had understood that this movement produced a controllable visible result he wanted to try again on his own, and repeated it over and over again."[6] This was a magnificent moment in time when the hand of the ape was guided by that of the human. Considered the "Picasso among great apes" by Wilson,[7] the young chimpanzee Congo produced approximately four hundred paintings (with fingers or with a paintbrush) and drawings (about half of his output), in six years, starting in 1956. This diligent practice obviously plays an important role in the process of emerging expertise. Thus "he gradually learned how to hold it between his thumb, index and middle fingers, which allows the greatest precision in drawing."[8] Over time, his concentration had intensified to a level where he could paint for about an hour: "As the weeks passed, he gained in confidence and every coloured line or blob was placed exactly where he wanted it, with hardly any hesitation [. . .] Nothing would interrupt him until he was satisfied with the balance of his picture."[9] Both his manual precision and his technical know-how were thus honed over time. He even made a change to his gestures which may, at first sight, seem trivial, but for Morris is crucial: instead of drawing his shapes fanned out, from the top of the paper toward the bottom, he unexpectedly inversed the movement, and started at the bottom before progressing up toward the top of the page. Furthermore, Congo began to focus increasingly on issues of composition,[10] clearly revealing a genuine taste for symmetry and balance. He thus spontaneously explored different procedures and a large repertoire of pictorial signs, repeating some patterns and then varying them. Moreover, he would only agree to relinquish his works when he felt that the paintings were complete: "Congo would put down his brush or hold it out to Morris when he wanted to move on to another sheet. To remove the page before the end or to insist that he continue with a painting judged by him to be complete would cause considerable annoyance."[11]

Painting Apes

Christine, a chimpanzee belonging to photographer Lilo Hess,[12] a natural history enthusiast, loved colors: "Her enthusiasm knew no bounds when she was presented with a paintbox. She adored handling the colour and spreading it over a sheet of paper with her fingers."[13] But apes not only display marked sensitivity to colors and shapes: they also clearly pay attention to lines and to thickness of the stroke.

The graphic and pictorial activity of apes also responds to an intrinsic logic. It is neither disordered nor chaotic. For Morris, these great apes are displaying deliberate acts of creation. Study of their activities allows one to gather evidence for aesthetic qualities, styles that are peculiar to each individual, an elementary sense of composition, a meticulous choice of colors, and a determination to accomplish the work: ". . . Bella, a gentle female whose concentration during painting sessions was absolute; she never lost her temper, even when her sweets were taken from her, until one day her keeper wanted to remove her painting materials at an inopportune moment."[14] The chimpanzees spend a great deal of time painting, showing sustained concentration, which they lack when engaged in other activities. Schiller describes their motivation as "profound" and "spontaneous." This is clearly supported by films and photographs of these apes at work. Their style evolves over the years and their productions display pronounced inter-individual differences. Morris considers that the apes display a certain desire for order in preparing their compositions. For the Belgian philosopher and art historian Thierry Lenain, they carefully organize their painting, but according to principles of perturbation, rather than of order: "In order to develop the capacity for graphic symbolisation in an ape a means would have to be found of diminishing the animal's instinctive enjoyment of the game of disruption."[15] It is also possible that these primates have a need to *feel* the pictorial field, the colors, the paintbrush, to engage in hand-to-hand combat with the paper, which could lead them to proceed by layers, each new layer systematically perturbing the previous one. Finally, it should be noted that these painting apes are not rewarded or encouraged. In 1957, Morris organized a sale of these creations, in the heyday of American abstract expressionism. The exhibition was well received. One of the "works of art" was sent to Jackson Pollock, who praised its aesthetic qualities. Some of Congo's paintings were sold at auction in Chicago, New York, and London, where prices reached 12,000 British pounds (in 2005). The presence of these ape paintings in salerooms, in museums and exhibition galleries, admits them culturally and economically into the art world. Yet, this still does not mean that the players in this market are truly ready to consider them *works of art.* Wolfgang Köhler, Paul Schiller, and various primatologists also report that apes possess a disposition and a spontaneous liking for painting and graphic design, something that Lenain confirms. Roger Fouts proclaims that the chimpanzees at his center love "making art." They even give their paintings titles, titles that they indicate using sign language. For instance, Washoe gives the name "Electric Hot Red" to a "work" in which red dominates. The male Dar has a really special manner of using the pigments: he eats them. His older sister, Tatu, would rather give up a meal than leave a painting unfinished. Ally displays a very energetic style, as well as a surprising similarity, from a technical viewpoint, with Jackson Pollock's "action painting." Moja is known as the first great ape to have produced representational paintings. She drew a round shape with an orange marker, and then explained in sign language that it represented a cherry. She apparently also painted a flower, and a bird as well. The gorilla Koko, raised by Francine Patterson in California, portrayed the same subject, drawing arched and plunging lines on the paper, suggesting the curvature of flight. In 1981, Roger Fouts set up an exhibition of the works of four of his talking chimpanzees. The works of Washoe's group are shown to an art critic, who considers that the apes could be heirs of Pollock. One must of course be cautious when making such comparisons with various artistic movements. Indeed, detractors of the avant-garde of the time did actually use the monkey paintings to vilify the works of some artists, by equating them to the "scribbles" of the great apes and by describing the works as "monkeying about." When, in 1964, the Swedish journalist Ake Dacke Axelsson exhibited the works of the French painter Pierre Brassau in the Christinae gallery in Gothenburg, these were greeted with effusive praise. But, in fact, Peter, a four-year-old chimpanzee, had produced the paintings! The critics were wrong. Recruited as an instrument in the ensuing quarrel, the output of the painting apes becomes suspect and the topic of their artistic abilities is abandoned.

When Elephants Paint . . .

Primates in zoos also show a particular taste for painting and even provide themselves with the means to exercise their talents. At Antwerp Zoo, the gorilla Victoria coated her finger with bird droppings and drew a harmonious composition of signs and lines on the window of her enclosure. Similarly, a young mangabey at the Ménagerie of the Jardin des Plantes used his own droppings to draw figures resembling suns on the glass panes of his cage. In the same primate house, the orangutans inscribe patterns of crisscrossing lines on the metallic panels of their cage, as soon as they can get hold of a screw or some other sharp instrument. Various species have shown some form of pictorial talent to differing degrees: cats, sea lions, horses, beluga whales, seals, dogs, and elephants. As far as elephants are concerned, we need not dwell too long on the videos devoted to these pachyderms in Thailand, which draw elephants as if producing self-portraits, and are actually trained to do so. These performances are linked to a much larger enterprise designed to preserve Asian elephants, and notably directed by the Russian artists Vitaly Komar and Alexander Melamid, founders of a painting school for elephants.[16] These elephants are retired as they are no longer allowed to work on the land, because of massive deforestation and the abandonment of traditional agriculture, and have therefore been "recycled" by taking on a function in the entertainment industry. The sites where their performances take place are part of the touristic circuits and their "works" are sold both in the sanctuaries and online.[17] Some of the "painting elephants" make "art" in an unconstrained manner and any project that they undertake allows them to perform an activity that interests them, while for others (such as those involved in the famous self-portraits) painting is akin to training for a circus act. On the other hand, a very interesting book[18] devoted to the drawing activity of the elephant Siri was published. Through her, the two authors carry out an investigation of the aesthetic sense of pachyderms. At the time, James Ehmann was a science journalist and David Gucwa an animal technician at Burnet Park Zoo, in Syracuse (New York state). Gucwa was taking care of Siri,[19] an Asian elephant weighing almost four tons and with a shoulder height of some eight feet, and noticed that the female would often trace lines with a stone on the ground in her interior enclosure. Her movements were clearly deliberate. In 1980, he decided to devote his breaks and free time to drawing sessions with Siri, who was about twelve years old at the time. During these sessions, he provided her with paper, paints, paintbrushes, or pencils. Without ever being encouraged, or rewarded, she freely drew dozens of compositions on sheets of paper: intertwined lines with a graphite stick, thick lines or more compact shapes with a paintbrush. She produced these in 20 to 30 seconds, sometimes pausing to examine her drawing. She would hold the graphite or brush with her trunk, which is a very flexible, precise, and sensitive instrument. The keeper reports that Siri from the outset showed considerable delicacy and intelligence in her creations. Nonetheless, her pictorial technique evolved considerably over the course of the sessions. The authors then tried to show her works to painters and art critics. Some refused to meet them, notably Desmond Morris, while others met with them and examined the drawings of their protégée with interest. They were often impressed by Siri's productions. Willem de Kooning and his wife, Elaine (also an artist), found that they are undeniably endowed with style, a good dose of determination, and considerable originality. de Kooning exclaimed: "That's a damned talented elephant."[20]

Others considered that Siri's productions were "poetic," delicate, elegant, and even "artistic." They were in no way accidental or random occurrences. Some specialists nevertheless compared them with children's drawings, without taking into account the sophistication of the work, the impressive delicacy of the strokes, and the particular sense for lines. The elephant still lives in Syracuse, and Gucwa declares that he learned more from Siri than Siri learned from him. Other elephants, such Ruby at Phoenix Zoo, Maharani Indira at the Metro Toronto Zoo, the entire group at the Washington Park Zoo in Portland (Oregon), all paint, and they do not do this to attract tourists. Zoos, however, quickly saw the benefits that could be gained by selling these paintings. The elephants display impressive capacities for focusing their attention, great precision, and rhythm. And even a certain grace. They know what they are doing and evidently enjoy doing it. And then there is John Eisenberg (University of Florida, Gainesville), who observed "wild" elephants in Sri Lanka who seemed to be drawing in the sand with their trunks. The great apes are thus not the only animals to exhibit aesthetic sensibility. In a book written for the general public, but rarely listed in his bibliography, the French philosopher Etienne Souriau, a specialist in aesthetics, shows how much the *aesthetic act* seems to be linked to "impulses stemming from the depths of life."[21] I follow with a few, particularly unsettling, examples.

François-Bernard Mâche's "Ornitho-musicology"

François-Bernard Mâche is a composer, musicologist, archeologist, writer, and member of the Académie des Beaux-Arts, currently holder of the Xenakis chair. He was a student of Messiaen, who was himself also fascinated by birdsong[22]: "Messiaen is correct in saying that many birds are not only virtuosos but artists, above all in their territorial songs."[23] Four to five thousand species of bird sing, and among these about 200 to 300 produce melodies that are particularly interesting for musicians. A leader of the so-called naturalistic aesthetic, Mâche defends an aesthetic approach to these songs, one founded on the hypothesis of gratuitous sound production. As the founder of zoomusicology (1983), he considers that the phenomenon of music is more widespread across the living world than we tend to think.[24] He shows, with backing from sonograms, that some birds are composers of true musical creations. The continuum extending from these songs to music has been acknowledged for centuries by many musicans, including Handel, Mozart, and Vivaldi. Darwin himself noted that birds have a *taste for the beautiful,* as did the American biologist Craig Wallace. For Mâche, it is clear that these avian musicians possess a great artistic sense, complex *musical cognition*, and considerable imagination. Their melodies testify to this, being non-stereotypical, inventive, individualized, and variable across space and time. Starting in the 1960s, Mâche accordingly integrates the vocal creations of several species, which he transcribes meticulously, into his scores. His transcriptions, together with syntactic and statistical analysis of the songs, allow him to reveal striking resemblances with diverse human musical elements: rhythms, ornamentations and accelerations, alternation between verses and choruses, syntaxes, and polyphonic organization. The similarities are far too numerous for anyone to believe they are illusory, or pure effects of convergence. In fact, the ethnomusicologist George Herzog arrived at the same conclusions,[25] as he considered that one cannot really find any fundamental differences between human music and birdsong. On the other hand, because they are generally not musicians, few biologists can really appreciate the musical qualities of these melodies.[26] They focus above all on the *function* of the songs: delimitation of territorial boundaries, vocal signature transmitted to other males, role in female choice, calling to the partner. *On the contrary*, Mâche shows that birds' songs are by no means solely governed by functionality. Their richness and their variety go far beyond what would be required to fulfill the functional needs described by scientists. In fact, the composer was surprised by the quality of some of the melodies in fall and winter, as these are even more sophisticated than their emissions in spring. Indeed, these songs would seem to escape, at least partially, from their functionalist shackles, as they occur during periods when sexual selection is not a prominent feature. What is more, it is common for males to continue to sing even after females have been fertilized. Thus, according to Mâche, an aesthetic function is present in songbirds. They are thus revealed as being both musicians and creators: "Everything indeed happens as if they had at their disposal simultaneously the many pieces of a rich repertoire, a corpus of rules governing their sequence and a generous amount of individual imagination allowing instant inventiveness."[27] Like Mâche, Rosemary Jellis considers that the melodies of several species exhibit levels of exuberance and refinement that are not necessary for the functions they are supposed to serve.[28]

Playful Activities, Fantasies, and the Pleasure of Sounds

The American aesthetic philosopher Charles Hartshorne[29] and the Hungarian musicologist Peter Szöke[30] also defend the aesthetic dimension of bird songs, which vary greatly between different individuals. In *Le sens artistique des animaux*, Etienne Souriau writes that the bird "names itself and announces its presence, with all its might."[31] Indeed, birds abandon themselves to a playful activity expressing a true physical euphoria, a form of intrinsic pleasure linked to the song: "It would seem that the bird experiences a specific pleasure in singing, and that its zeal is due at least as much to this pleasure as to the possible result of an amorous success."[32] Souriau adds: "Indications of intrinsic pleasure derived from sound are not lacking in birds."[33] Moreover, it is the birds with the subtlest songs that spend most of their time vocalizing, and seem to be least constrained by the functional shackles attached to them.[34] Likewise, they seem to experience a Funktionslust, "a pleasure in doing what they know how to do well." In fact, their enthusiasm is reinforced by the excitement provoked by their own melodies, as these exert a conditioning effect on their performance. During a private interview, François-Bernard Mâche confirmed to me that the notion of Funktionslust seemed to him to be perfectly appropriate. Moreover, even if Souriau never really explicitly refers to this notion himself, it still seems to be clearly present in his writings: "During this action, to which he devotes himself more and more completely, powers of aestheticism arise and are developed, and it is not difficult to enumerate the main principles. There is, first of all, the free momentum to which he abandons himself. We noted previously this kind of 'bodily enthusiasm' [like that] observed in a dog excited by the tumult of the waves. Yet here is a type of enthusiasm in which some factors are undeniably of a psychic nature. . . . To all this, one must add the excitation that the bird derives from its own song, as a performance, through perceiving it; akin in fact to the excitation that it receives from the song of a rival that responds to it from afar in the night. . . . Because at the center of this there is indeed a creation. The song of the accomplished singing bird is not simply the 'ordinary' product of a pure act of emission . . . It is by destination, by function, a transmission towards a real or virtual listener. It is a thing that exists to be heard."[35] Last but not least, many bird species are capable of *transposing* and *imitating* various sounds, noises, and melodies of individuals of their own or other species. This, for Mâche, indicates a real interest in sound forms and an aptitude to manipulate them. For this reason, the zoomusicologist Hollis Taylor became interested in the pied butcherbird, a "prodigy of imitation."[36] Other species also deploy remarkable musical aptitudes—for example, male and female gibbons, which excel at performing duets,[37] or cetaceans, with their impressive songs that are being studied with increasing intensity.[38] Some musicians and conductors have accordingly become fascinated by the musical skills of marine mammals—for example, Jim Nollman (who notably tried to create musical dialogues with killer whales by using drums)[39] or Alan Hovahaness (who integrated whale song into his compositions).[40]

Hawaian Dolphins and Their Bubble Rings

Another striking example also testifies to the creativity of cetaceans, more specifically of the bottlenose dolphin (*Tursiops truncatus*). In the 1990s, biologists Ken Marten, Karim Shariff, Suchi Psarakos, and Don White[41] were working on various research topics with some dolphins at Hawaii's Sea Life Park at Makapuu Beach, on the island of Oahu.[42] They were notably interested in testing the dolphins' intelligence and self-awareness, by means of touch screens. The cetaceans did not participate in these experiments unless they wanted to, and were never rewarded. The researchers very swiftly noticed that the dolphins engaged in a fascinating activity: making bubble rings. To achieve this, they produced whirlpools with their flippers and then breathed out air through these eddies with their blowhole. In this way they fashioned circles,[43] spirals, tori, vortices, and helices. Thus, these dolphins are not only intelligent but were also revealed as being extraordinarily imaginative. Building on a basic technique, they managed to multiply the shapes they created, by varying their blowing patterns, the motion of their flippers, and the position of the body. Once the bubble rings had been shaped, the dolphins played with them, and amused themselves by moving their own bodies or the bubbles, bursting them or making them slide along their bodies, in a delicate aquatic ballet. The adult male Kaiko'o actually managed to merge two distinct circles. Stabilizing the bubble and maintaining its shape depends on appropriate control of the eddy and the pressures exerted in different places. The dolphins thus implement complex knowledge concerned with the physical properties of water. They hence detect, understand, and manipulate certain subtle dimensions of their environment, using their sonar to localize the invisible whirlpool where they will blow the bubble. The researchers underline the fact that this is a leisure activity, which does not seem to be connected to any functional form of behavior. Nevertheless, this "tradition" of blowing bubble rings spread to several members of the group, with the less experienced dolphins observing the experts. Making bubbles properly and perfecting them demand months of practice. The dolphins that have not yet mastered these techniques (some eight individuals out of seventeen) carefully watch the other group members, then try to repeat what they have seen. It is fascinating to note the degree of sophistication of the skills involved in the context of these activities, which seem totally gratuitous. Studies conducted at other dolphinariums have revealed that the activity of the bubble blowers of the island of Oahu is unique and that this group genuinely displays a "tradition of bubble rings" that is lacking from other groups. Remember that the young of these marine mammals spend four to five years with the mother, a period during which they watch her closely and constantly, and acquire different types of know-how through imitation. In fact, dolphins are known to imitate one another. It seems that this is a privileged means of learning for these species, as is the case for great apes. Moreover, another research team has shown that the bottlenose dolphins of Hawaii are able to control the qualities of the bubble rings they create, and anticipate their movements during their games with them, by supervising them both visually and through echolocation. They confirm that the production of these circles requires extensive training, implying an understanding of physical causality, with the cetaceans mastering the necessary knowledge for manipulating the characteristics of their aquatic environment. The creator of Project Delphis, Don White, sees in these activities a dimension that has been ignored by science. Describing the dolphins' realizations as kinetic sculptures, he considers that they are indeed creating art. According to him, their motivation is both playful and aesthetic: "For myself, I do consider these rings to be 'art': the creation and observation of artifacts by a nonhuman mind, with no use other than entertainment and aesthetics. One must be constantly wary not to anthropomorphize the actions of other species—to treat them as though they were human. But after watching a dolphin create one of these kinetic sculptures—observe it from many angles and then destroy it with a bite—it seems a long leap of logic to ascribe any other motive."[44]

In the Kingdom of the Bowerbirds

In the nineteenth century, a number of naturalists, among them William Jardine and Prideaux John Selby, described strange constructions.[45] These first observers report that they resemble, on a small scale, indigenous huts with a well-maintained garden.[46] Some even assume that they are actually indigenous children's dollhouses, as their appearance mirrors that of human constructions. John Gould,[47] professor of ecology and a specialist of animal behavior at Princeton, undertook to study these astounding structures in the Australian bush. In fact, these gardens represent the nuptial stages of *bowerbirds,* members of the family Ptilonorhynchidae. These passerines are the closest relatives of the birds of paradise, and include eight genera and about twenty species.[48] They are found in New Guinea and on neighboring islands, as well as in some parts of Australia. At least twelve of these species construct bowers of two types: *avenue-type bowers* or *maypole bowers.* In the first type, the avenue is constructed with two walls consisting of thousands of branches, twigs and various grasses. Rendered rigid by weaving the plant materials together, the walls demarcate an area for parading, with the dance floor located at the entrance. The other type of bower is produced around a central pillar (the maypole), generally a tree trunk. It is arranged in the shape of a "circular passage" carpeted with moss around the central mast, which is furnished with a rainproof roof. Construction of both types of bower demands the use of large quantities of materials. They are further decorated with shiny elements, flowers, feathers, berries, leaves, small stones, brightly colored objects, fragments of bleached bone, the skulls of small animals, jawbones, and shells,[49] the exoskeletons of crustaceans, nails, bivalve mollusk shells, insect wing cases, pieces of metal, fragments of glass, discarded plastic items, and shiny human artifacts (for instance, car keys, bottle caps, coins, rifle shells, teaspoons, or even metal cups). Up to 120 types of different objects have been recorded, and these are carefully selected, collected, and employed, according to the species, individual preferences, and local "culture." Several pounds of stones are found strewn along the avenues, which are sometimes over three feet long.

"Its Passion for Flowers and Gardens Is a Sign of Good Taste and Refined Manners"[50]

The construction of these bowers was explained, in part,[51] by the rather drab colors of the male's plumage in some of these species. These constructions (simultaneously an extension of their body and courtship behavior), would thus serve to compensate for their lack of physical appeal, helping them to attract a mate.[52] However, this hypothesis was not retained: different species of bowerbirds are brightly colored and they nevertheless construct magnificent bowers. Only males construct the bowers, which never serve as a shelter for the future brood: the females build their own bowl-shaped nests, and rear their offspring in them, without any help from the father. Gould explains that bowerbirds have few competitors for food resources.[53] They thus have more spare time at their disposal for building their stage: they devote about 80% of their activity to this task. Furthermore, they show themselves to be extremely demanding, even dismantling their construction if they are not satisfied with it. Moreover, they renovate the bowers constantly and add new elements. The first to admire the vault of the Vogelkop bowerbird, or Vogelkop gardener bowerbird (*Amblyornis inornata*), was Odoardo Beccari in 1872, who wrote enthusiastically that the bird's "passion for flowers and gardens is a sign of good taste and refined manners." It is true that this bird is an avid collector of colors and objects. The male adores red and orange, and embellishes the enormous structure that has taken years to weave around a trunk, with repeated improvements, using flowers, berries, and mushrooms in these hues. He is generally surrounded by about seven competitors living within earshot. Every male fashions his own individualized bower, one choosing darker colors, another using manure, and a third adding original objects to the construction. In contrast, the male satin bowerbird (*Ptilonorhynchus violaceus*) is attracted by elements that are colored blue, notably various accessories found near human dwellings (bottle caps, pieces of plastic, etc.). As in other species, for example, the yellow-breasted bowerbird, he paints the interior walls of his bower using a fibrous piece of wood anointed with a mixture composed of berry pulp and saliva, sometimes with the addition of wood ash. Also, he may occasionally coat his breast with blue. He can push his sense for detail to an extreme, possibly taking a parrot feather, also blue, in his beak during his nuptial parade, and making the iridescent reflections play in the light. Indeed, the visual effect thus produced assumes a crucial importance during the courtship process. The satin bowerbird adds all kinds of sounds to his display, including metallic noises. If he is particularly efficient, he might successfully fertilize about thirty females over the course of a season. The bower of the great bowerbird (*Chlamydera nuchalis*), the largest representative of the group (with a body length of around fourteen inches), is particularly impressive. The documentary maker Heinz Sielmann took an opportunity to examine one of its bowers. It contained between 5,000 and 12,000 components, including 500 bleached kangaroo bones and 300 pale yellow snail shells. Altogether, the collection of materials gathered together weighed between twelve and twenty-five pounds. In fact, this bird deliberately chooses ornaments that highlight its own plumage and manipulates the decorations of its bower to create interesting contrasts. Moreover, the spotted bowerbird (*Ptilonorhynchus [Chlamydera] maculata*) is the only animal species known to cultivate a plant for a reason that is not linked to food. This bird grows flowering plants known as "potato bush" (*Solanum ellipticum*) in its nuptial area. Bearing violet flowers and green fruits, plants of this species help to aggrandize the structure of the bower.[54] Finally, for many naturalists, the structures of the streaked bowerbird (*Amblyornis subalaris*) are the most beautiful ever to be erected by a bird. The central tower of its bower, the trunk of a tree fern, is enhanced with magnificent decorations, and appears like a mosaic decorated with multicolored floral ornamentations, berries, leaves, and beetle wings. The stage is carpeted with mosses that are dotted with fruits or red flowers that are renewed constantly.

Other birds build structures that are fascinating because of the techniques used. Weaverbirds fashion their nests in the shape of a cone, by using a dozen different types of knot, just like real clothesmakers. The European penduline tit builds a shelter that resembles the finest and most symmetrical basketwork. That of the Eurasian oystercatcher is furnished with shiny pebbles. The tailorbird sews one leaf to another with its beak using a very fine stalk as its thread, and then lines these two leaves with feathers, downy plant material, and cotton. Finally, caciques weave nests that measure up to 20 inches across. However, the bowerbirds are the only ones to construct such extravagant and complex nuptial stages, and to show real concern for ornamentation and an obvious *sense of beauty*.

The Sense of Beauty[55]

For Gould, bowerbirds in fact possess a sense of beauty and are hence able to evaluate the loveliness of their bower. Their bower is also judged by the females, according to the number and originality of its component elements, the symmetry of its layout, the rigidity of its "walls," its design plans, and the freshness of the berries and flowers with which it is decorated.[56] The bowers are inspected while the male is absent, each female returning to those that met her approval, so as to be present at the nuptial parade, but without copulating at the end of it all. A female then takes about a week to build her nest, before returning to be courted in the bowers that she preferred. The bower is hence only part of the male's plan of action: added to it are his movements and leaps, his dance moves, his courtship postures, his plumage, as well as his various calls (whistling, chirping, or imitations of the sounds of local species). During his parade, he fluffs his feathers and his frills, and throws brightly colored objects in the direction of the female, as if they were "presents." Everything is undertaken so that the whole package is as elaborate and impressive as possible, from an *aesthetic* point of view.

The female chooses one male, a single male, and then copulates with him, after having adopted a position indicating that she accepts impregnation by him. The male's display then becomes less vigorous, because he has to mount his lady-love at the right moment, as otherwise mating would be less effective. The suitor can in fact display violence during the parade, as the ardor of his display contributes to his power to convince. The barriers of branches, erected by our bower builder, serve to protect the females, which simultaneously seek intensity and security. The more experienced females (and hence those that are less easily scared) seem to attach greater importance to the performance, while the younger ones are more impressed by the bowers. The degree of complexity of the structures, their ornamentations, and the performance of the male testify to the quality of the lover and constitute as many ways of showing himself superior to his rivals, to different audiences: females, inexperienced males, or directly competing males. Each suitor must use enormous amounts of energy, considerable skill, and experience to build and maintain his stage, and protect it from his competitors. The number of fruits, objects, flowers, is also a reflection of the bird's capacity to repel his rivals. He reveals his competitive ability as well as his reproductive potential. When researchers added berries to a structure, the male removed them until the quantity was the same as at the outset. He hence aims to maintain a balance between increasing the value of his stage and limiting attacks by his competitors, which might destroy the bower if it were flamboyantly embellished with too many ornaments. In addition, there are "fashions" among bowerbirds: those males that want to seduce a greater number of females need to create styles that run counter to the prevailing trends in a given area.[57] Local and regional differences emerge, shaped by the choices of the females being courted.

Optical Illusions

According to Gould, males are able to adopt the female perspective, very subtly, in order to construct attractive bowers. The high degree of aesthetic refinement displayed by suitors is undoubtedly linked to their capacity to anticipate the responsiveness of those they court, assessing the reliability of their evaluation through the number that accept copulation. This evaluation of the females' judgment amplifies (like a positive feedback loop) their drive to construct the most beautiful bowers and to improve them constantly. Gould argues that this process results in selection for particular cognitive abilities. Thanks to the capacity of male bowerbirds to see things from the perspective of a lady-love, they go even further to please her: they use optical illusions. In fact, researchers John Endler, Lorna Endler, Natalie Doerr, and Laura Kelley[58] have shown that the great bowerbird renders his bower more attractive by manipulating the perspective: by doing this his ability to seduce a coveted female is increased.[59] The entrance and exit of his nuptial stage (oriented north to south) are decorated with diverse objects, which the bowerbird accumulates while selecting them carefully according to their size, from the smallest (at the entrance, the female side) to the largest (in the background in relation to the position of the female). Entering this avenue from the south side, the female thus sees it from constant angles. Despite being larger, the stones that are furthest away seem to have the same size as those that are closer. In this way, the bower appears smaller than it is, and, thanks to the optical illusion, the male seems more impressive by contrast. This arrangement is deliberate. When the researchers disrupt the assembly and arrange the elements in the opposite direction, the males put them back in their initial order within a few days. It seems that this manipulation of the visual perspective (called "forced perspective" or "altered perspective") does actually influence the choices made by females. Given their degree of complexity, erecting bowers and perfecting the nuptial parade demand a long apprenticeship. To make progress, males must practice intensively. Juveniles train together on rudimentary stages, which they construct and dismantle. It takes them at least seven years before they are able to construct satisfactory stages. This long process of learning is partly made possible by the life expectancy of the bowerbirds, the longest of any passerine family. These young males take their inspiration from the bowers of the most experienced males, whose constructions they inspect during their absence. Competition is intense: when a male starts building his own bower, he risks seeing it pillaged or even destroyed by his older rivals. Charles Darwin himself also described these birds in *The Descent of Man and Selection in Relation to Sex* (1871). He had visited Australia during his voyage on the *Beagle*. For him, it is clear that the females feel genuine admiration for the males and their bowers. He even thinks that they *derive some kind of pleasure from the sight* when inspecting them. In fact, it would seem that the male's performance triggers a striking multisensory experience in the females, a "range of modes of experiencing that add themselves to vision—including rhythm. Each perceived quality brings to the expression a web of inherent connections, which can extend as far as including all available modes (hearing, touch, smell): it is a *plenum* of experiences."[60]

Artistic Mannerism?

In *A Thousand Plateaus*,[61] Gilles Deleuze and Félix Guattari also focus in detail on bowerbirds. They demonstrate that *Scenopoietes dentirostris* (tooth-billed bowerbird) is sensitive to color contrasts. Leaves that have been previously severed and dropped to the ground are turned so that their paler undersides contrast with the soil: "inversion produces a matter of expression."[62] The bowerbird uses these as landmarks, points of reference that he maintains, replacing faded elements with fresh ones. This scene is framed by the walls woven out of plants. The assemblage is completed by his movements and by his remarkable song, which he emits while standing on a branch situated above his arena. He also imitates the melodies of other birds and various additional sounds. The nuptial stage represents an open arena for the actions of the male in his search for mates. For Deleuze and Guattari, this is how a territorial apparatus is composed, consisting of sounds, colors, plant elements, earth, and movements. The bird "makes himself visible and sonorous at the same time."[63] It is the environment itself that provides different expressive materials, which the bowerbird endows with meaning. The two authors add that this "assemblage" cannot be reduced to the functions linked to territoriality or sexual selection. Expressiveness precedes it: "The territory is not primary in relation to the qualitative mark; it is the mark that makes the territory. Functions in a territory are not primary; they presuppose a territory-producing expressiveness. In this sense, the territory, and the functions performed within it, are products of territorialization. Territorialization is an act of rhythm that has become expressive, or of milieu components that have become qualitative."[64] For them, functionality is hence *secondary* compared with the expressive forces that participate in territorialization and which construct the territory. The richness of the ornaments present in the bowers greatly surpasses what is needed to serve the function of reproduction. The exuberance and the complexity of these nuptial stages defy the mechanistic and functionalist approach that is current in scientific circles. Complicity with the components of the environment leads, in this way, into expressive forms that are subtler than would be expected and which testify to an extraordinary *sense of beauty* in bowerbirds. The biologist and ethologist Karl von Frisch, a bee specialist, who received the Nobel Prize in Medicine along with Konrad Lorenz and Nikolaus Tinbergen in 1973, concurs with this sense when he describes the activities of the bowerbird: "Every time the bird returns from one of his collecting forays, he studies the overall color effect. He seems to wonder how he could improve on it and at once sets out to do so. He picks up a flower in his beak, places it into the mosaic, and retreats to an optimum viewing distance. He behaves exactly like a painter critically reviewing his own canvas. He paints with flowers; that is the only way I can put it. A yellow orchid does not seem to him to be in the right place. He moves it slightly to the left and puts it between some blue flowers. With his head on one side he then contemplates the general effect once more, and seems satisfied."[65] How can we think of sexual selection without assigning a sensitivity that we would call "aesthetic" to the participating animals? The degree of sophistication of this sensitivity is so advanced in bowerbirds that Hansell uses the phrase "artistic mannerism."

The Aesthetic Senses of Primates, Cetaceans, and Birds[66]

Many primatologists are convinced that great apes possess aesthetic sensitivity[67]: some adorn themselves with leaves or flowers, or use ochre as "make-up," and then look at their reflections in rivers. One female uses mud as lipstick. And rehabilitant orangutans are fascinated by soap bubbles.

They blow into the water and observe the resulting motion. For Anne Russon, orangutans have a sense of order and of symmetry, universal criteria of *beauty*. For instance, they make bundles with long sticks of the same size, which are impressively parallel. According to the neurobiologist Jaak Panksepp[68] (Washington State University), the pleasure felt when faced with these manifestations of "beauty" would be connected to a primordial emotional system, which is found in both humans and animals: he calls this the *blue-ribbon emotion system*. He defines this system as being related to four fundamental elements, which are connected to precise zones of the brain: seeking, rage, fear, and panic. Functioning as a general platform for the expression of numerous basic processes, the *seeking system* conditions the necessary motivation for approaching or avoiding an object in the environment, as well as the possibility of anticipating what positive or negative elements this object could provide.[69] When this center is stimulated in humans, they feel the same kinds of emotions as other species, notably a basic impulse to investigate their environment, and to find elements that have meaning for them. The first stage of this process is an attraction to novelty. Yet, when an unknown element appears in our world, this prompts a response. Two possibilities then present themselves, fear on the one hand ("fear, that fundamental and almost constant basis of accommodation in the wild"[70]) and appetite on the other ("appetite that makes him tingle, while approaching step by step, the cat ready to jump on its prey"[71]): either the object represents a danger that one must flee from, or it is a new object that the organism tries to explore or acquire. It is of course this second, highly stimulating, type of experience that is sought. Indeed, the *seeking system* can be linked to positive emotions that are measurable and very pleasant, and notably come in the form of desire, of expectation, or of curiosity.[72] Furthermore, it has been clearly shown that the hippocampus and the *ventral tegmental area* of the brain contain circuits for the detection of novelty, which has the effect of activating a mechanism linked to dopamine.[73] In *Animals Make us Human*[74] (a book in which she mentions the work of Panksepp), Temple Grandin also emphasizes that animals create opportunities to stimulate the *seeking system* for themselves, to such a degree that scientists think that the system functions like a pleasure center within the brain.[75] Such emotional conditioning is tightly linked to the aesthetic contributions, which are inherent to natural selection.[76]

Ready-made . . .

The multiple, heterogeneous elements of the environment are thus converted into sources of pleasure, which is potentially innovative. Manipulated by mammals or birds, these resources can become expressive materials, as demonstrated by Deleuze and Guattari: "Territorial marks are readymades. And what is called *art brut* is not at all pathological or primitive; it is merely this constitution, this freeing, of matters of expression in the movement of territoriality: the base or ground of art. Take anything and make it a matter of expression."[77] This process of *becoming expressive* is thus mobilized by different species, with objects in the environment being implemented in games of sexual selection, converted to territorializing landmarks, or transmuted into artifacts that are potentially devoid of any functionality. Gathering, transforming, and staging environmental elements generates a certain *pleasure,* from a functionalist perspective: an unexpected *by-product,* as it were. Clearly present in knot-tying apes, the musical songbirds, bubble-blowing dolphins, and bowerbirds, this principle of pleasure is at the core of different activities where the products exceed the function that they are presumed to serve. These products are not solely linked to selection, but become expression, "expressive excess in comparison with the norm."[78] The neo-Darwinian notion of adaptation is thus reevaluated. As explained by Brian Massumi, American philosopher and translator of Deleuze, "there is, shrouded in its instrumentality, a push towards the excess suggesting another natural aestheticism—or another nature of aestheticism. It is hardly a coincidence that Deleuze and Guattari base their theory of the animal on a theory of art."[79] Massumi enriches the issues with the question of supernormal signals. For instance, human or animal babies show various features, including a head and large eyes that are disproportionate in comparison with the body (resulting in what is called cuteness, a property that is used abundantly in cartoons, notably by Walt Disney). On the other hand, the parents are sometimes equipped with reinforced identifying signs. A famous study was conducted by Tinbergen on the European herring gull, which presents a contrast between the white head, the yellow color of the beak, and the red spot that it bears. This contrast constitutes a stimulus, which drives the chick to peck at the base of the beak to incite the adult to disgorge some food. These signals are described as *supernormal* by the founder of comparative ethology, Niko Tinbergen, and as *supra-normal* by Massumi: "The supra-normal in fact indicates a plasticity in the limits of the natural, a process of deformation, that is to say of transformation—a process that generates an internal drive and operates through normal behavior towards a surplus of attested value."[80] This extension of plasticity renders the experience all the more intense, the additional dimension creating *experiential qualities* at the individual level. In connection with the question of the *seeking system,* the need to face unexpected conditions has the outcome that stereotypical behaviors are not sufficient. In environments that are constantly changing, the *powers of improvisation* of an individual are in fact essential for its survival. Perpetual change elicits the response of *transformation* of the organism, an opening of the genetic program towards a dimension linked to intuition, improvisation, and thought. Territorialization is exactly one of the modalities that allows "as an activity in the literal sense, the surge of the element of improvisation, which was already at work in the instinctive act."[81] Moreover, the *pleasure of doing what one knows how to do well,* the Funktionslust, supports these practices and functions like a drive for learning, as well as, at another level, a powerful evolutionary catalyst. *Pleasure* is hence truly at the heart of a system, which leads to different forms of expertise. This is equally decisive under the action of sexual selection. In bowerbirds, the females become excellent critics over time. Their task, however, is arduous, as they have to judge about ten bowers, composed of hundreds of sundry elements. Hansell argues that they in fact do not compare the bowers themselves, but the *pleasure felt* contemplating each one. This process is in fact simpler than performing an analysis taking into account the sum of multiple components of each nuptial area, demanding memorization and comparison (the Art School Hypothesis).[82]

The Immediacy of Creation

Thus transported by the emergence of pleasure and expertise, the performing animal seeks, selects, collects, manipulates, experiments, invents. Like a large number of living creatures it is interested in the elements of its environment, seeking new objects in it, and then using these in its own creations, in a sensitive choreography with the objects of the world. It abandons itself completely to its "task," plunges into the *immediacy of creation*, and devotes itself to this wholeheartedly with a controlled agitation, abandoning itself in the urgency of the moment. It is important to note that it is the *moment itself* of realization that *matters*. Intentionality is not located in the project of *realizing a work*, but occurs in this *immediacy of creation* and in the desire to *do*. In any event, the totally disconcerting objects that emerge from this process cannot be simply described as crafts, in the sense that apes, dolphins, and birds are not aiming at producing something that is strictly useful. They are not trying to manufacture chairs. They detach themselves instead from pure functionalism, going far beyond that, even if just in the imaginativeness and exuberance of their artifacts. Nests could simply permit the apes to sleep; they do not need to be beautiful, or bestow magnificent panoramic views. Yet they attain an astonishing degree of refinement through the choice of colors and plants, their simplicity, the selection of twigs of the same length, utilization of specific plants, applying the teeth to the edges of the bed, and the choice of view. Knotted objects, nests, paintings, songs, bowers, and bubbles without doubt show the qualities that we describe as "aesthetic" and, sometimes even, as true "artistic affectation." They manifest an intentionality, a *presence*, a *presence of self*, a sense of beauty and a presence, sufficient to mark them out as exceptional, without us needing to invoke the notion of *art*. All the more so because definition of this term is obscure and subject to constant revision, varying greatly from one era to another, and from one horizon to another, and dependent on fragile criteria for demarcation of what is a "work of art," in the current artistic context. Some animal creations reveal to us, according to *our* criteria, what exists in different places and at levels that are imperceptible to us, at the core of life itself. The aesthetic fact, which we have classed among our highest virtues, is more widespread than we thought. Yet, because it does not fit the logic of measurement or mathematical models, or enter the experimental paradigm that presides over any serious study of the "animal," it has long been ignored. The topic of aesthetic phenomena in animals has remained largely unexplored, either with scientific tools (poorly fitted for this field of investigation) or with tools of the social sciences[83] (immediately disqualified by natural scientists, particularly because of their monopoly of the study of the living). These manifestations have hence been totally ignored or reduced, restricted to the limited dimensions that can be linked to a biologizing perspective and to a functionalist concept of nature.

An Immense Vital Symphony

Characterized by a kind of gratuitousness and going beyond strict functionality, the behaviors described in this chapter seem to escape from these views. They further allow us to think of a world that is common to different species and to our own. The continuity between animal and human worlds does not belittle humanity, but enables us to encounter non-human creatures with greater generosity,[84] prompting us to pay attention to their existence and their expressions. In fact, the insights presented here speak in favor of the idea, defended by the art historian Pierre Sterckx,[85] of a "zone of indiscernibility," situated between nature and culture, humanity and animality, a zone in which art, creation, and pleasure are all involved. However, it is not of major interest to ask whether bubble-blowing dolphins, knot-tying apes, musical songbirds, painting apes, or bowerbirds are artists. What counts above all else is that their productions, which are simultaneously inventive, creative, frivolous, or exuberant, represent *manifestations of self* for individuals, testifying to their openness to the world, in ways that belong to them alone. The magnificent simplicity of Deedee's sheaf of long grass, adorned with three evenly distributed blue ribbons, the colorful installation constructed by Wattana, the exquisite passerine songs, the lightness of bubbles combined with the virtuosity of Kaiko'o or Tinkerbell, united in a subtle aquatic ballet, the aesthetic sense of the bowerbirds . . . All these creations are bearers of *life itself*.[86] They seed their environment ("For Masson, as for Pollock, to remake the world one must seed it"[87]), and reveal different forms of expertise, all in service of a vibrant *presence in the world*. We are, without a doubt, biased by the current definition of what we call "art" in the present day: a process that is characterized by inscription in the history of Western art, marked by an obligatory reflexivity of the author in relation to her condition, being both a human being and an artist, as well as a "meta" reflection concerning art. In any event, a *discourse* of the creator is necessary (the work sometimes being reduced to this *discourse* alone), and without this discourse (whether it is produced by the creator or various commentators), the creator is not considered to be an artist. From this viewpoint, the animal has nothing to say.[88] He remains mute. But does this definition not mislead us into underestimating some of the incredibly refined accomplishments of certain animals? "Is it truly blasphemy to think that art has a cosmic basis and that one finds in nature great founding powers with which it is congeneric?"[89] For Deleuze and Guattari, art is not a solely human privilege.[90] "Not only does art not wait for human beings to begin, but we may ask if art ever appears among human beings, except under artificial and protracted conditions."[91] What we call "art," without being able to provide a precise definition, is manifested in a multitude of ways, from the most humble to the most sophisticated: "Aesthetic acts present in abundance in nature."[92] There is a "speciation" of these art forms. Even if the uniqueness of the works produced by humans should not be underestimated,[93] that of the creations of dolphins, great apes, and birds should still not be ignored. Belonging to a form of *art of daily life*, of *art of life*,[94] their productions testify to a shared sensitivity, a drive of life and a power of desire, a common world and a common openness to the world. The songs of the avian musicians echo the noises of nature: rain drops, the crackle of leaves, rustling. Everything becomes melody. Kaiko'o makes use of his knowledge of whirlpools, of the physical properties of water and air, to invent kinetic sculptures. Congo paints. Deedee fashions strange and beautiful objects. The bowerbirds devise luxuriant gardens. Serving as openings to cosmic rhythms, these ephemeral productions, these activities that are both joyful and solemn, link us to the creatures and objects of the world, and refer us to an essential community, at the heart of an "immense vital symphony formed around thousands of presences: flowers, herbs, small worms, insects, birds, small forest quadrupeds, the whole, where, through a thousand links, everything answers everything. There, every being has its place and its role. And the bird that announces its presence, which listens to the answering calls, which defines its space, plays its minute ceremonial role within the immense morning ritual of the forest."[95]

EPILOGUE

Animals, in order to belong to our world, must of course enter into it to some degree. They must consent, even if only a little, to our way of life, they must tolerate it . . .

—RAINER MARIA RILKE AND BALTHUS, *Mitsou. Histoire d'un chat*, 59

Captive great apes have done much more than *consent* to enter our worlds: they have shown extreme goodwill in creating an existence for themselves within them. They do not even seem to hold it against us that we have obliged them to do so. By entering our zoos, they have also slipped into our lives. They are thus present at the crossroads of several entangled histories,[1] with their *personal story* mingling with that of their ape or human partners, inscribing them in a *social history* as well as in a specific *cultural history*. Nevertheless, we still continue to confine them to a single history: their *natural history*, reducing them to representatives of their species and products of their phylogenetic history. Yet, as this account has shown, the behaviors of great apes, eminently social and flexible, cannot be generalized at the species level. It is essential to think of the great apes as individuals actively connected to the "dispositive" (in the

Foucauldian sense), in which they occur. In the case of Wattana, this environment is that of an urban zoo established at the end of the eighteenth century, and consisting of heterogeneous elements: employees of the Ménagerie, listed buildings, limited surfaces, internal zoo rules, a management subordinated to that of the Museum, visitors, distribution of tea at five o'clock, breeding plans, direct contacts between the keepers and the apes, etc. The life of the young female Wattana is thus tightly connected to a very particular *cultural history*, linked to the objects and traditions adopted by the humans that surround her. This history is made up of habits, skills, and ways of living in the world, and is also nourished by the culture peculiar to the orangutans at the Ménagerie. For all that, this does not mean that Wattana's *natural* or *phylogenetic* history does not count for anything. It is not suggested that these dimensions should be in opposition but rather be allowed to be complementary, and that one should be open to possibilities that we tend to ignore as soon as we consider the "animal." Nevertheless, some of these dimensions are not in the domain of studies by biologists, in spite of the essential role they play. In fact, the scientific method aims to reduce the number of research parameters in order to be able to control for their influence. But, even if accounting for these different historical trajectories does complicate the economics of behaviors, it allows us to shine new light on great apes.

"We, All, Are Trying to Make a Living": A Personal History[2]

It is difficult for us to admit that creatures other than *human* beings might have a biography. Yet, great apes are born, grow up, meet others, form friendships, travel from place to place, breed, create families, age, and die. They all have very definite personalities, developing different character traits, preferences, varied interests, and particular skills. Fortunately, ethologists are now beginning to open themselves up to this long-ignored dimension of individualization. For her part, Wattana had a very chaotic start to her life. She was abandoned by her mother, and then had to leave the keepers in Antwerp who had taken care of her for the first three months of her life. After arriving at the nursery in Stuttgart, she finds a replacement mother in Margot Federer. But she is separated from her and from her playmates at the age of two-and-a-half. She leaves for Paris. At the Ménagerie of the Jardin des Plantes, she experiences further separations: her half-brother is sent to Hungary, several of her favorite keepers move to other positions, and her daughter, Lingga, is sent to a zoo in the United Kingdom. On June 3, 2008, Wattana herself is transferred to Apenheul Nature Park in Apeldoorn (Netherlands). All necessary precautions are taken to guarantee that she is as comfortable as possible. Her favorite keeper, Christelle, personally settles the ape into the transport crate. Once in the air-conditioned truck, Wattana is able to communicate during the whole trip with the accompanying animal technician Stéphanie. The female orangutan spits, whines, and negotiates to obtain treats. She does not seem stressed. After a journey of over six hours, Wattana arrives at the Dutch zoo. Christelle was only able to return to her after her working day had finished. It is then too late to visit her, but the next morning, as soon as the zoo is open, she is in front of her cage. Wattana immediately goes to her and reaches out to her with her hands. Christelle feeds her, and ensures that she really has everything that she needs. Her new keeper, Wilma, had already gathered information on Wattana's personality, her behaviors, her life story, before spending a few days with the orangutan in Paris, to allow them both to become accustomed to each other. It is important for the "reference" keeper to symbolically pass on duties to the new keeper in charge of the transferred individual. On the first day, Christelle takes care of Wattana as usual, but with the new keeper present. The next day, the two animal technicians take care of the orangutan together. On the third day, the Dutch keeper takes care of everything, under the careful supervision of Christelle, who then has to leave the Netherlands, with a heavy heart. She leaves behind a very dear friend, to whom she had become attached over several years. She knows that she will have only a few opportunities to visit her former charge. From now on, Wattana is thus cut off from her previous companions, and most notably from Tubo, the father of her daughter. Beyond these physical separations she must get to know new places, as well as keepers and a language that she does not know. She is also forced to make a place for herself within a new social group. At first, she shares the cage of an experienced forty-eight-year-old male. The keepers hope that Karl, born in the wild in 1961, will teach her various social skills. At the time, Wattana is in visual contact with several mother orangutans, including one who had raised two young ones, and another who had raised three. She hence has the opportunity to observe mothering females of her own species. It is thought that this might help her learn to take care of her future offspring.

Wattana's mother, Ralphina, lived at Apenheul between July 20, 2001, and January 26, 2008, the day of her death. It would take a few months before our Parisian orangutan had the opportunity to see the one that had given birth to her once again. On the other hand, another member of her family has been living at Apenheul since April 2002, her sister Binti, whom she is finally able to meet. Wattana is then kept in an enclosure occupied by the young male Amos.[3] On February 22, 2010, she gives birth to a son, Kawan, son of Karl. Greatly surprising all those who know her, she behaves very gently with her child, and this time is more comfortable in the role of a mother. It should, however, be emphasized that she had only then reached the age that is typical for orangutans to give birth to their first baby (she was over fourteen years old). By contrast, she was not even ten years old when Lingga was born. Furthermore, in the meantime she had been able to observe the mothering behavior of several female orangutans. Despite similar opportunities, also spending much time around other mothers that were attentive to their offspring, her mother Ralfina gave birth to four babies and did not take care of a single one.

Surrounded by humans from the day she was born, Wattana remained profoundly connected to their world. When I visited Wattana in Apenheul, in the summer of 2012, I gave her a short ribbon. She tied it around the wire mesh of her enclosure. She had not changed her habits and was still intent on keeping a close watch on her human companions. One evening, while she is spending time on one of the outside islands, she seems frightened by something close to the entrance to the inside quarters. She refuses to come back in. It seems that she received an electric shock, as there is an electric barrier to prevent the apes from leaving the enclosure. Although this incident occurs in summer, her keeper Leo Hulsker is nonetheless worried about his charge's comfort. He decides to approach her, in order to bring her a blanket and a little food. The young female accepts the fruits and then

Figure E.1 Wattana with her son Kawan, courtesy of Apenheul Primate Park, Apeldoorn (the Netherlands), 2010.

embraces Leo, slipping with him under the blanket, and carefully wrapping them both up together!

Nowadays, Wattana shares her enclosure with her son and Silvia, a female who is almost fifty years old and takes care of Kawan as if she were his grandmother. Wattana continues to devise her own opportunities to practice the skills she has learned from humans. For example, she unties the tension cords of the hammocks and uses them to tie knots. She also tears up pieces of cloth from old clothes provided by the keepers and ties them to the rings of her cage. She clearly retains her keen interest in knots. She also appreciates the opportunity to paint that her keepers provide for her during the cold season.

Families, Circles, Communities: A Social History

During their lives, great apes "socialize" with different partners: members of their own species, keepers, visitors, guards, veterinarians, researchers. They

Figure E.2 The twins of Vandu, Wattana's brother, courtesy of Hong Kong Zoological and Botanical Gardens, Hong Kong, 2011.

live in an extended social group, partly consisting of the circle made up by their family (when they are fortunate enough not to have been separated), as well as other apes that live in captivity alongside them. They are also part of the circle of human beings who gravitate around them, forming "human-ape communities." Closely linked to the *life history* of primates, this *social history* is similarly embellished with encounters, separations, losses, and periods of mourning. After her enforced separation from Margot Federer, Wattana also lost her most precious support: her half-brother, Vandu, transferred to Hungary in 2001. Later, in January 2010, this male orangutan left Europe to reside at the zoo in Hong Kong. On July 8, 2011, his companion Raba gave birth to twins, one female and one male.[4] Twin births are extremely rare in great apes. In the wild, it would actually be very difficult for a female to look after two young ones. Since orangutans have been kept in captivity, only four twin births have been recorded. Vandu, Raba, Wan Wan, and Wah Wah accordingly achieved true star status in Hong Kong.

Incidentally, the two sons of Nénette and Solok are now no longer with us: Dayu died on October 12, 2007, and Tubo on August 5, 2011. So there are now no living descendants of the large male, as the son and daughter born in Leipzig both died, one at birth and the other, Datu, at less than a year old. Wattana's enclosure companion for ten years, Nénette, the matriarch of the ape house, lost all her four sons, the two first ones, Doudou and Mawa, having died at the ages of eleven and twenty-six. She does, however, have two granddaughters: Lingga (daughter of her son Tubo with Wattana) and Surya (daughter of Mawa and Mota).

At Antwerp Zoo, Wattana's father, Tuan, was separated from his partner, Astrid, after thirteen years of cohabitation. At the end of the month of June 2007, he was moved to Chester Zoo, in Great Britain, while Astrid was sent to the zoo in Osnabruck, in Germany. Tuan then had two more offspring, with two different females: a son, Iznee, with Sarikei (twenty-five years old), and a daughter, Latifah, born about a month later, with the young Leia (thirteen years old). They live with one of the oldest captive orangutans, Martha, who is about to reach the venerable age of sixty.

Wattana's daughter, Lingga, left the Ménagerie of the Jardin des Plantes to go to England at the age of two, in July 2007. The search for a suitable adoptive mother had led to Wareham, in Dorset. It is essential for young ones that have been reared "by hand" to be integrated into a "normal" social group as soon as possible, so that they can acquire the habits and customs of their species. Lingga now lives at Monkey World. The small female hence had to leave the animal technicians that had fed and pampered her. She shares the new nursery with six other orangutans.

A small tribe of females of different ages currently reigns over the Ménagerie: Nénette (about forty-six years old), Theodora (twenty-six), and her offspring Tamu (ten). Theodora and Tamu were transferred to Paris from Twycross Zoo on November 28, 2007. They were introduced to Nénette's enclosure on June 25, 2008, and seem to be getting on well. The keepers reported that Theodora and Tamu can make knots.

With captive orangutans, family members are frequently dispersed to the four corners of the earth, so they rarely have the opportunity to live together or at least see each other, even if only very infrequently, as they might have been able to do if they had lived in the forests of Borneo. Within their group, they have positive relationships with some individuals and keep their distance from others. Moreover, they interact more frequently with humans than with members of their own species. Playing the role of replacement parents and foster fathers, the keepers represent extremely important social partners for great apes, as well as being role models and *charismatic leaders* that are carefully watched and imitated. The apes establish selective affinities with some of their keepers. They then share with them a kind of *between-species friendship*, which is characterized by very strong bonds of trust. Despite all this, they are still often separated, whenever individual apes are transferred from zoo to zoo, or when there are personnel changes among animal technicians.

From Nests to Knots: A Natural History

Thanks to the opportunities presented by human environments and the possibilities that these create, captive orangutans in zoos are able to practice skill types that have developed during their slow coevolution with various components of humid tropical forests. Because of the flexibility and plasticity of these primates, their natural abilities and technical skills are refined in a world that differs drastically from their original environment. In particular, the gestures involved in nest construction and their fibroconstructive techniques find new applications in the weaving, wrapping, and interlacing that apes notably employ to fashion hammocks, swings, or beds for the night. Some apes manage to tie knots, reinvent the principle of the lever, or clean windows. Others apply their mechanical skills and dexterity to loosen the bolts of their cage, to disconnect electrical systems,[5] or to tear away small pieces of metal from the enclosure structures in order to use them to pick locks. Their inquisitiveness is evident in their attraction to anything novel, and in the keen interest they show in the birds, mice, and worms that frequent their territory. Their ingenuity, their imagination, their predilection for solitary pastimes, their patience, and their ability to plan actions in advance are put to use during a wide range of activities, allowing them to fashion objects that are occasionally extraordinary. However, they do not employ the full range of their skills because of the impoverished kind of environment that is typical of most zoos.

Encounters between Humans and Great Apes: A Cultural History

Wattana is a participant in the long history of encounters between humans and apes. This is a history that begins in ancient times, when humans were intrigued when first being confronted with these creatures, which resemble us in physical appearance, anatomy, and behavior. This interest reached a particular turning point with the voyages of exploration organized by a few major European nations, starting in the fifteenth century. The first great apes arrived in Europe two centuries later. Grafted onto this history is that of the animal parks, symbols of the splendor of those in power since antiquity. An annex of the Muséum d'Histoire Naturelle, the Ménagerie du Jardin des Plantes, is created, after the French Revolution, and after the pillaging and the destruction of the Royal Ménagerie at Versailles, an emblem of the greatness of the French monarchs. In the nineteenth century, zoological gardens multiply in the West. From this point forward, they are linked to the colonial history of Western nations. Great apes, which should never have been subjected to imprisonment, flooded in. Moreover, initiated by scholars of the Enlightenment, another history—that of research focusing on primates—commences at the end of the eighteenth century. This discipline adopts the name *primatology* in 1941. The spaces where humans and great apes coexist are diverse: comparative psychology centers, biomedical laboratories, private houses, homes of former expatriates, sanctuaries, and zoos. In places where primates live in close proximity to human beings, there is an exchange of habits, abilities, technical skills, and fragments of ethos. Some individuals learn to tie knots, move about bipedally, barter, paint, use iPads, watch television, or drink tea. Thanks to this encounter with human cultural objects, apes also engage in activities that reveal a certain aesthetic sensibility. Moreover, apes that tie knots, paint, or create varied and extravagant artifacts display a style that is individually unique. However, even though captive great apes appropriate diverse elements of human culture, they also share in the different skills and living styles of the group of their own species to which they belong in any zoo. The keepers observe all of these phenomena every day. As a result, their expertise is essential for piecing together knowledge regarding captive great apes.

Re-enchantment of the World?

The way that great apes inhabit the spaces we have assigned them to is thus nourished by different histories, be they *biographical, social, cultural, natural,* or *phylogenetic*: Wattana has enabled an encounter between, among others, the Empress Joséphine, the dolphins of Hawaii, the city of Paris, the professors at the Museum, the first orangutan to arrive in Europe, knots, the animal technicians to which she was linked, Messiaen, Nénette, mirror neurons, the tropical forests of Indonesia, and their simian inhabitants. At the center of this confluence, a strange little female orangutan was able to unite around her keepers, veterinarians, visitors, specialists in animal cognition, and philosophers. Her serious looks, her eagerness to indulge in activities she is passionate about, her willingness to acquire our knowledge, the precision of her gestures, her dignity, her profound attachment to humans, all signal an ethos, a way of *being,* in the world that touches us deeply. Every great ape similarly has a face, a life, a history. Wattana, Tuan, Tubo, Dayu, Solok, Deedee, Nénette, Lingga, Teak, and Astrid re-enchant the world, not due to their presence in our zoos (how could one be enchanted by such imprisonment?), but because they infuse our lives with a poetic dimension that blends the power of *life itself* with an aesthetic sense and an art of everyday life, and through the fleeting traces they leave in our world, hallmarks of their *presence*. The "means of transforming life are provided by life itself."[6]

ACKNOWLEDGMENTS

I express my most sincere gratitude to Lidia Breda, Jean-François Lamunière, as well as Benoîte Mourot and Maryse Djian, and the staff at Éditions Payot & Rivages.

Thanks from the bottom of my heart to Dominique Lestel, my accomplice in this adventure of knot-tying apes, as well as to my writing accomplice, Lauren Herzfeld.

I would also like to thank Christelle Hano, Gérard and Danielle Dousseau, Émilie Gouverneur, Franck Simonnet, Géraldine Pothet, Stéphanie Labbé, Valérie Martinez, and the keepers at the Ménagerie of the Jardin des Plantes. They generously shared their experience, their knowledge, their memories, their passion. Without them, Wattana's biography would never have seen the light of day. The animal technician Margot Federer vividly portrayed for me Wattana's childhood and everyday life at the nursery of the Wilhelma Zoological-Botanical Garden in Stuttgart. To her, I extend my warmest thanks. I am also grateful to the keepers at the zoos of Planckendael and Antwerp, as well as Leo Hulsker, Cora Schout, and Bianca Klein of Apenheul Primate Park, in the Netherlands. I cannot overstress the importance of the knowledge of the keepers, as they are able to describe the behavior of their protégés with incredible precision. Living in close proximity with the great apes, they collect observations of incomparable value and show a capacity for fine judgment that is rarely equaled.

Heartfelt thanks go to Jacques Rigoulet (Muséum National d'Histoire Naturelle, Paris), who is also fascinated by the history of the Ménagerie of the Jardin des Plantes, where he previously served as director. My exchanges with Richard Burkhardt (University of Illinois), science historian and leading expert on the history of the Ménagerie, were also very fruitful.

Outstanding gratitude is also due to Linda Van Elsacker (Antwerp Zoo), Jeroen Stevens (Planckendael Zoo), Hilde Vervaecke, Lorin Milk (Tampa's Lowry Park Zoo), Nannette Driver-Ruiz (Fresno Chaffee Zoo), Lyn Myer (Fresno Chaffee Zoo), Jane Anne Franklin (Louisville Zoo), as well to all those who were kind enough to reply to my investigations through e-mail.

Lyn Miles shared her experiences concerning Chantek. She also allowed me to take photographs of several artifacts that had been produced by this orangutan, whom we visited together at Atlanta Zoo. To her, too, I extend my warmest thanks.

Thanks to Cécile Chaussey who drew my attention to Wattana's knot-tying ability, and to Bernadette Bensaude-Vincent and Isabelle Stengers, with whom I had many fruitful discussions while writing an initial version of this study.

Film director Florence Gaillard spliced together all the sequences that I had shot during the experiments carried out at the Ménagerie. Thanks to her, three films devoted to showing Wattana's knot-tying were produced. I thank her for the tremendous amount of work she invested.

Thanks also to Henry, Lauren, Nathan, and Elisabeth Herzfeld, my most loyal supporters.

Last but not least, I wish to pay tribute to Wattana, who has given me so much, and also to her father, Tuan. I cherish particularly fond memories of the moments that we spent together face-to-face. At the time, I was unaware that Tuan was the father of the one who would acquire such importance in my eyes. I was lucky enough to meet representatives of four generations within this family. I first got to know Tuan, and then encountered Wattana, and later her daughter, Lingga, in 2005. Finally, I was also able to meet Wattana's grandfather, Maias, before his death on November 14, 2005. The keepers at Cologne Zoo, where the large male spent all his time in captivity (over thirty-six years), were the ones who told me that he knew how to tie knots.

NOTES

Introduction

1. Jane Goodall, *The Chimpanzees of Gombe: Patterns of Behavior* (Cambridge, MA: Belknap Press, 1986), chapter 4, "Who's Who," 60–78.

2. Dian Fossey, *Gorillas in the Mist* (Boston: Houghton Mifflin Company, 1983).

3. Also known as anthropoid apes because they have features resembling those of human beings.

4. A reference work, specific to each great ape species, listing information on births, transfers, and deaths of all captive individuals.

5. See, for example, William T. Hornaday, ed., *The Minds and Manners of Wild Animals: A Book of Personal Observations* (New York: Scribner's, 1922).

6. See, for example, Henri Coupin, *Singes et singeries* (Paris: Vuibert & Nony, 1907) or, more recently, Eugene Linden, *The Parrot's Lament and Other Tales of Animal Intrigue, Intelligence, and Ingenuity* (New York: Plume, 2000) and Eugene Linden, *The Octopus and the Orangutan* (New York: Dutton, 2002).

7. The term "anthropomorphic" is used for a creature having human form.

8. Gibbons belong to the family Hylobatidae, which is also part of the superfamily Hominoidea. Gibbons have long been known as "lesser apes," contrasting with the other species known as "great apes." Gibbons are, however, very different from the great apes.

9. Belonging to the primate order, the Hominoidea excluding gibbons are characterized by lack of a tail, large body size, a face with a real nose, and a large brain volume. They consist of four genera: *Pan* (chimpanzees and bonobos), *Gorilla* (gorillas), *Pongo* (orangutans), and *Homo* (humans). In earlier classifi-

cations, the great apes were placed in the family Pongidae, separated from the Hominidae (containing only modern and extinct humans).

10. Phylogenetic: pertaining to patterns of genetic relationship between species.

11. Various preserved remains provide evidence that orangutans also lived in other parts of the Asian continent, notably in China and Thailand.

12. And not 10,000 years as suggested by some authors, even quite recently. See work by Devin Locke et al., "Comparative and Demographic Analysis of Orang-utan Genomes," *Nature* 469 (2011): 529–533, and the article "Orangutan DNA More Diverse than Human's, Remarkably Stable through the Ages," available on the site of Washington University in St. Louis (School of Medicine), January 24, 2011: http://news.wustl.edu/news/Pages/21774.aspx

13. In Sumatran orangutans, both females and males typically spend all their time in the trees. In contrast, big male Bornean orangutans, especially older males, occasionally move down to ground level. This is much rarer in females, which are vulnerable due to their smaller body size.

14. Kalimantan: Indonesian part of the island of Borneo, consisting of four provinces.

15. These are generally the darkest of all orangutans: their fur is a dark black-brown color. Nevertheless, there are occasionally large differences between individuals within the same subspecies.

16. Source: International Union for Conservation of Nature (IUCN)—http://cms.iucn.org/

17. http://www.iucnredlist.org/details/17975/0

18. http://www.iucnredlist.org/details/39780/0

19. While Heidegger certainly does not consider himself a follower of Descartes (he uses him instead as his prime target, both for his mechanical philosophy and for his *cogito*), he does remain Cartesian as far as animals are concerned.

Chapter One

1. This part of the book outlines historical aspects of the Ménagerie of the Jardin des Plantes. For his valuable inputs, I am greatly indebted to the science historian Richard W. Burkhardt (University of Illinois), a leading specialist in the history of the Ménagerie.

2. See: Jacques-Henri Bernardin de Saint-Pierre (Intendant au Jardin national des plantes et de son Cabinet d'histoire naturelle), *Mémoire sur la nécessité de joindre une ménagerie au Jardin national des plantes de Paris* (Paris: De l'imprimerie de Didot le jeune, 1792; 63 pages).

3. Aubin-Louis Millin, Philippe Pinel, Alexandre Brongniart, *Rapport fait à la Société d'Histoire Naturelle de Paris sur la nécessité d'établir une ménagerie* (Paris: Imprimerie de Boileau, 1792).

4. Only the zoological garden in Schönbrunn at the imperial summer residence near Vienna is more ancient. This institution has, however, been changed profoundly over time, even if the menagerie retains its original appearance. Furthermore, the garden was not opened to the public until 1779, more than 25 years after its creation in 1752 by Emperor Francis I of Habsburg-Lorraine.

5. During the French Revolution, countless birds disappeared from the royal menagerie. Maintaining the cages and feeding the many occupants was extremely expensive. The royal gardens were considered a symbol of absolutism, and the Jacobin revolutionaries and fur traders helped themselves to numerous primates, deer, and various birds.

6. Louis-François Jauffret, *Voyage au Jardin des Plantes, contenant la description des galeries d'histoire naturelle, des serres où sont renfermés les arbrisseaux étrangers, de la partie du jardin appelée l'école de botanique; avec l'histoire des deux éléphans, et celle des autres animaux de la ménagerie nationale* (Paris: De l'Imprimerie de Ch. Houel, 1797–1798 [An VI de la République]).

7. Richard W. Burkhardt, "La Ménagerie et la vie du Muséum," in *Le Muséum au premier siècle de son histoire*, ed. Claude Blanckaert et al. (Paris: Éditions du Muséum national d'Histoire naturelle, 1997), 491.

8. The dromedary that carried Napoleon during the French campaign in Egypt finished its days in the Ménagerie.

9. It was not the first of its kind to arrive in France: the great naturalist Buffon lived with the famous young female chimpanzee Jocko, from 1742. She died in 1744 and was autopsied by Daubenton. She is still kept in the collections of the Museum as a stuffed specimen.

10. Eric Baratay and Elisabeth Hardouin-Fugier, *Zoo: A History of Zoological Gardens in the West* (London: Reaktion Books, 2004), 79.

11. Taxonomy: science of naming and classifying organisms.

12. See Donna Haraway, "Teddy Bear Patriarchy: Taxidermy in the Garden of Eden, New York City, 1908–1936," *Social Text* 11 (1984): 19–64 and Donna Haraway, "Teddy Bear Patriarchy: Taxidermy in the Garden of Eden, New York City, 1908–1936," in *Modest Witness@Second Millennium. Femaleman Meets Oncomouse: Feminism and Technoscience*, ed. Donna Haraway (New York: Routledge, 1997).

13. The apes were given to the Ménagerie by donors from all walks of life: ship captains, missionaries, private individuals, governors, ship doctors, colonial and navy officers, merchants, medical institutes (notably the Pasteur Institutes in Kindia and in Paris), scientists in charge of expeditions for the Museum (Urbain, Rode), and researchers (Voronoff).

14. Eric Baratay, "Belles captives: une histoire des zoos du côté des bêtes," in *Beauté animale* (catalogue de l'exposition), ed. Emmanuelle Héran (Paris: Réunion des musées nationaux/Grand Palais, 2012), 200.

15. Edouard Bourdelle and Paul Rode, "Note à propos d'un jeune orang (*Pongo pygmaeus* Hoppius) né à la ménagerie du Muséum" (Paris: *Bulletin du Museum*, 2e série, t. 3, 1931, 475–78).

16. This zoo was reopened in April 2012 after more than two-and-a-half years of major renovation work.

17. See Chris Herzfeld, "L'invention du bonobo," *Bulletin d'Histoire et d'Epistémologie des Sciences de la Vie* 14 (2007): 139–62.

18. The last two bears at the Ménagerie, Pacha and Louise, departed for Touari on April 28, 2004.

19. After yet another refurbishment, these bear pits now serve as an enclosure for the red pandas and binturongs.

20. See Shirley C. Strum, *Almost Human: A Journey into the World of Baboons* (New York: Random House, 1987).

21. Orangutans are said to be "semi-solitary" and to exhibit "fission-fusion grouping." Contrary to the other anthropoid apes, they do not live in groups in their natural habitat (apart from a few populations in Sumatra).

22. Great apes can be linked by some forms of friendship just like humans.

23. From the 1930s, and particularly between 1960 and 1990, several programs for

teaching human language to great apes were set up across the USA. The apes learned to communicate with their instructors either through sign language or by using icons arrayed on symbol boards, presented as plastic elements or displayed on computer screens.

24. Gérard Dousseau relates how this was the reason given to him by one of the people in charge at the time, who added, "We are not at the circus" (personal communication, April 2004).

25. Sometimes the concept of "dispositif" (in French) is translated as "apparatus." See Jeffrey Bussolini, "What Is a Dispositive?" *Foucault Studies* 10 (2010): 85–107.

26. See for example the excellent *Great Apes Enrichment Project* website: www.gaep.eu

27. Including mangos, mangosteens, lychees, rambutans, or durians.

28. It is difficult to estimate ages of captured great apes accurately. In the *Studbooks*, the margin of error for age estimation is taken as approximately two years.

29. Named to honour the Ménagerie's director, François Doumenge.

30. Bernadette Bresard, "Problème de l'asymétrie cérébrale chez les anthropoïdes. Contribution à l'étude de la phylogenèse des processus cognitifs" (PhD diss., Université Pierre and Marie Curie [Paris 6], 1983).

31. Marie-Christine Lacour, "Communications interspécifiques (chimpanze, orang outan, homme). Exploitation d'un recueil de séquences d'expressions faciales: Préparation à l'informatisation" (master's thesis, Université Paris Diderot [Paris 7], 1983).

32. Captive orangutans are so adept at integrating into their human surroundings that some individuals may be able to understand several hundred words.

33. See Dominique Lestel, *Les amis de mes amis* (Paris: Éditions du Seuil, 2007).

34. At the Tanjung Putting National Park in Indonesia, orangutans hold a little bundle of leaves to their bodies in their nests during sleep, very much like human children might clutch dolls. Scientists consider this behavior to be peculiar to that particular group.

Chapter Two

1. According to scientists, female orangutans can become fertile at roughly six to eleven years of age, and attain sexual maturity between the ages of eight and 15. Sexual maturity in males is reached around the age of 12. The menstrual cycle lasts between 22 and 30 days, while the gestation period is from 237 to 270 days (i.e., between eight and nine months).

2. Furthermore, as living conditions have detoriated so markedly in the natural habitat due to commercial exploitation and resulting massive deforestation, mainly because of palm oil production, several observers have noted accelerated rates of development in young orangutans.

3. Leo Hulsker, personal communication, Apeldoorn, May 24, 2012.

4. In their natural habitat, female orangutans give birth to four babies on average. In captivity, some females have had up to eight offspring. Individual females have been reported to give birth when over 40 years old.

5. When a mother "rejects her infant" (using the time-honored expression), the tendency nowadays is to try to have the baby adopted by an individual of its own species.

6. They do not even know how to carry the baby in the correct fashion—that is, with the head uppermost.

7. Some sources mention ages of 58 years for males and 60 for females. However, one male from Ketambe (one of the oldest-established research sites in Indonesia) did reach the age of 60.

8. See Christophe Boesch, "Teaching among Wild Chimpanzees," *Animal Behaviour*

41 (1991): 530–32 and Roger Fouts, Alan Hirsch, and Deborah Fouts, "Cultural Transmission of a Human Language in a Chimpanzee Mother-Infant Relationship," in *Child Nurturance: Studies of Development in Nonhuman Primates,* ed. Hiram E. Fitzgerald, John A. Mullins, and Patricia Gage (New York: Plenum Press, 1982), 159–93.

9. I thank Jeroen Stevens (Royal Zoological Society of Antwerp), primatologist and bonobo specialist, for drawing my attention to this point.

10. This behavior of looking fixedly and intensely is known as "peering."

11. Not only are orangutans the gardeners of the forest, they are also excellent herbalists. For example, they use *Fordia splendissima* flowers against migraine and as a stimulant. They also use flowers of the *Melastoma* genus against diarrhea, as well as various plants against parasites such as malaria, and deposit *Campnospernum* twigs in their nest to repel mosquitos.

12. As regards progressing through the tress, their impressive arm span is a major advantage. This measures about seven-and-a-half feet in males.

13. It has been argued that, having been raised with several bonobos, Wattana demonstrated sexual behaviors characteristic of that species. For my part, I stress that I never observed this type of behavior, despite visiting Wattana at around the same time as the author of the article. See André Langaney, concerned "Watana, orang-outan immigrée," *Le Temps (Genève),* June 15, 2004. (Officially, Wattana is written with a double *t.* Every *Studbook* uses this spelling, so using an alternative version reflects inadequate documentation.)

14. Pierre Gay, *Des zoos pour quoi faire? Pour une nouvelle philosophie de la conservation* (Paris: Delachaux & Niestlé, 2005), 58–59.

15. Probes for extracting honey, branches for capturing insects, mittens and cushions made of plants, leafy branches used as fans, handfuls of leaves for cleaning faces, spoons for honey, leaves rolled up into the shape of straws for sucking up liquids that are out of reach, masturbation sticks, rods for "fishing" insects, etc.

16. For more information concerning cultural transmission via social apprenticeship: Michael Krützen, Erik Willems, and Carel van Schaik, "Culture and Geographic Variation in Orangutan Behavior," *Current Biology* 21 (2011): 1808–12.

17. Moreover, it needs to be underlined that the species is also becoming less and less solitary as a consequence of the drastic reduction in forest availability, inevitably leading to greater population densities.

18. See Chris Herzfeld, *Petite histoire des grands singes* (Paris: Editions du Seuil, 2012), chapter 4, "Des anthropoïdes qui se prennent pour des humains," 90–127.

19. *Galium,* a genus of plants belonging to the family Rubiaceae and related to the bedstraw found in North America, is a favored food of mountain gorillas.

20. Gilles Deleuze and Félix Guattari, *Kafka: Toward a Minor Literature* (Minnesota: University of Minnesota Press, 1986): "On the other hand, the imitation is only superficial, since it no longer concerns the reproduction of figures but the production of a continuum of intensities in a nonparallel and asymmetrical evolution where the man no less becomes an ape than the ape becomes a man. The act of becoming is a capturing, a possession, a plus value, but never a reproduction or an imitation."

21. See Gilles Deleuze and Félix Guattari, *A Thousand Plateaus: Capitalism and Schizophrenia* (London: Continuum, 1988), chapter 10, "Becoming-Intense, Becoming-Animal, Becoming-Imperceptible . . ." See also: Gilles Deleuze, *Francis Bacon: The Logic of Sensation* (London: Continuum, 2003) [French issue, p. 30].

22. Baratay, "Belles Captives," 196–209.

23. Whenever Solok was present, Nénette would always retreat to the same place

high up in the cage, thus creating a territory for herself. She was presumably attempting to avoid frequent contact (even visual) with the large male. This was the only way in which she could protect herself. Soon after Solok's death, she stopped spending all of her time on the mezzanine and lost a considerable amount of weight within a few months.

24. Experimental tests showed that male orangutans are seven times as strong as men. They are apparently able to lift 530 pounds without exerting themselves.

25. Valérie was obliged to leave the ape house after a work-related accident. Wattana was unable to see her again until several years later. She is usually extremely quiet and makes no noise at all. When she saw her former keeper, she immediately approached her and began emitting rather strange squeaking sounds. For all who witnessed their meeting, it was clear that the orangutan was expressing very strong emotions on being reunited with the woman who had been so important to her.

26. Christelle Hano is now deputy head keeper and responsible for the Monkey and Reptile departments.

27. Pp. 208–27.

28. Willie Smits uses a screening test approach: Gerd Schuster, Willie Smits, and Jay Ullal, *Thinkers of the Jungle: The Orangutan Report* (Königswinter: Ullmann/Tandem, 2008).

29. As far as males are concerned, the record seems to be just over six feet.

30. I provide these numbers as a guide, but they are estimates that can vary widely from one source to another (main sources: IUCN, Primate Info Net, Ecofac/CEE).

31. Eduard Tratz and Heinz Heck, "Der afrikanische Anthropoide "Bonobo": Eine neue Menschenaffengattung." *Säugetierkundliche Mitteilungen* 2 (1954): 97–101.

32. Gauri Pradhan (University of South Florida) and colleagues showed that male Sumatran orangutans (*Pongo abelii*) were able to delay the start of the second stage of maturity. They hence extended their adolescence, waiting to gain in strength and body weight to rival the large males who monopolize all females in their home ranges: Gauri R. Pradhan, Maria A. van Noordwijk, and Carel van Schaik, "A Model for the Evolution of Developmental Arrest in Male Orangutans," *American Journal of Physical Anthropology* 149 (2012): 18–25.

33. For Carel von Schaik, in addition to serving to demarcate territories of large males acoustically, these long calls also function to delay the start of the second stage of maturity in adolescent orangutans living in the vicinity, by triggering hormonal stress: Carel van Schaik, *Among Orangutans: Red Apes and the Rise of Human Culture* (Cambridge, MA: Belknap Press, 2004).

34. Four times a day on average. Calls are emitted more rarely during the night.

35. The facial disk of large males diminishes in size when young adults "usurp" them from power.

36. Solly Zuckerman, *The Social Life of Monkeys and Apes* (London: Kegan Paul, 1932).

37. Frans De Waal, *Chimpanzee Politics: Power and Sex among Apes* (Johns Hopkins University Press, 1982).

38. See work by Jeroen Stevens on the group of bonobos living at Planckendael Zoo, near Malines (Royal Zoological Society of Antwerp). Findings from this research have notably been evoked during the controversy between Ian Parker ("Swingers: Bonobos Are Celebrated as Peace-Loving, Matriarchal, and Sexually Liberated. Are They?" *New Yorker*, July 30, 2007) and Frans de Waal ("Bonobos, Left and Right: Primate Politics Heats Up Again as Liberals and Conservatives Spindoctor Science," *Skeptic*, August 8, 2007).

39. See, for example, Shirley Strum and Bruno Latour, "Redéfinir le lien social: des

babouins aux humains." In *Sociologie de la traduction, Textes fondateurs*, ed. Madeleine Akrich and Michel Callon (Paris: Presses de l'École des mines de Paris, 2006), 71–86.

40. It is more difficult in the ape house as metal plates, which are difficult to move, cover all the openings located high up. Nevertheless, Wattana and Tubo manage to move these decorative panels in order to uncover the slatted gratings at the tops of their cages. They then pass various objects (especially bits of paper) through the grating to visitors.

41. However, it is necessary to specify that in November 2013, a roof was installed on the outside cage. All the openings were then closed, in order to allow the orangutans to go out all year round, including during the winter time. Theodora and her daughter Tamu, two of the current residents at La Ménagerie, did not seem to appreciate this layout: Theodora showed her discontentment by banging on the windows. Their caregiver, Christelle Hano, thinks that this is linked to the fact that they no longer have contact with the public. The transmission of objects from humans to the great apes, and vice versa, has indeed become impossible.

42. See for example Jean Estebanez, "Le zoo comme dispositif spatial: mise en scène du monde et de la juste distance entre l'humain et l'animal," *L'Espace Géographique* 39 (2010): 172–79.

43. The apes are never punished but instead encouraged using a signal (the sound of a clicker or whistle, for example) or a reward, which communicates to them that they have accomplished the required action successfully.

44. This vocabulary, which is primarily one used by animal breeders, is applied in zoos. In fact, Gérard Dousseau points out that the best animal keepers come from a rural background. Nowadays, there are specialized training programs, yet it is still the case that keepers from the world of animal breeding stand out from the crowd. For Dousseau, keepers are above all other things specialists in animal breeding with a detailed knowledge of the requisite methods.

45. Tubo and Wattana were able to respond to more than seventeen requests.

Chapter Three

1. See the works of Tulpius, Tyson, Edwards, Buffon, Linné, Houttuin, Bontius, Daubenton, Camper.

2. The term was introduced by the Dutch medical doctor Jacob de Bondt, known as Bontius, who was sent to Batavia (now Jakarta) by the Dutch East India Company. Adapted from the term "orangutan" (orang meaning "person" and hutan, "forest"), used by the Javanese population, it is found in its Latin form as *Homo sylvestris* in the first edition of his *Historiae Naturalis* (1658).

3. Arnout Vosmaer, *Description de l'Espèce de Singe aussi singulier que très rare, nommé Orang-Outan, de l'Isle de Bornéo. Apporté vivant dans la Ménagerie de Son Altesse Sérénissime, Monseigneur le Prince d'Orange et de Nassau, Stad- houder Héréditaire, Gouverneur, Capitaine Général et Amiral des Provinces-Unies des Pais-Bas, Etc. Etc., Etc.* (Amsterdam: Chez Pierre Meijer, 1778).

4. Published in the *Magasin encyclopédique* (volume 1, number 3, pp. 451–53). Later, in 1798, they would publish their "Mémoire sur les orangs-outans" in the *Journal de Physique, de Chimie et d'Histoire naturelle* (III, 46, 185–91).

5. The skeleton examined by Cuvier and Geoffroy Saint-Hilaire is certainly the one that is listed under the inventory number #A 10722 in the comparative anatomy gallery of the Muséum National d'Histoire Naturelle. It is, however, unclear if it was the same skeletal specimen sent by von Wurmb to the Netherlands.

6. There are in fact males that reach a height of over five feet. The specimen must have been a juvenile or young adult, although the sex of the skeleton was not recorded.

7. The orangutan studied by Vosmaer had been sent from *Angkola* (Northern Sumatra), and this name was probably confused with *Angola*. It was hence thought that the animal came from Africa and was an ape related to the chimpanzee, maybe like the *pongo* described by Andrew Battell in the seventeenth century (which was most definitely a gorilla). Total confusion reigned.

8. "Pygmy" because their small size had surprised naturalists. These scholars had expected to discover beings that were intermediate between humans and animals, and hence of the same size as humans, fitting the image of hybrid creatures described by the Ancients.

9. This whole section refers to Giulio Barsanti's 1989 article: "L'orang-outan déclassé. Histoire du premier singe à hauteur d'homme (1780–1801)," *Bulletins et Mémoires de la Société d'anthropologie de Paris* 1 (1989): 67–104.

10. The "Petit Trianon" is the name of Queen Marie-Antoinette's private domain near which was her rustic retreat, the Hameau de la Reine, with a farmhouse and many animals.

11. See Charles-Maxime Catherinet de Villemarest, *Mémoires de Mademoiselle Avrillon: première femme de chambre de l'impératrice, sur la vie privée de Joséphine, sa famille et sa cour* (Paris: Garnier frères, 1896).

12. Hornaday, *Minds and Manners of Wild Animals*, 112.

13. William Hornaday, *Popular Official Guide to the New York Zoological Park* (New York: New York Zoological Society, 1923), 83, and Hornaday, *Minds and Manners of Wild Animals*, 112.

14. Personal communication, Lyn Miles, Atlanta, 2007.

15. *International Studbook of the Orangutan (Pongo pygmaeus, Pongo abelii). 2010*, coordinated by Megan Elder, Como Zoo, Minnesota, 4.

16. Hybrid orangutans are the product of crosses between orangutans of Borneo and of Sumatra. Nowadays, zoos try to avoid any such hybridizations, which would be impossible in nature between these geographically isolated species. In order to stop hybrid individuals from breeding, they are given contraceptives, or placed in enclosures where they cannot copulate with other orangutans. Nevertheless, such hybrid individuals are still occasionally born. In 2009, among the thirty-four births of anthropid apes in zoos, two were from such hybrid crosses. See the website of the *Orangutan Species Survival Plan* (plan initiated in 1982): http://www.orangutanssp.org/member-zoos.html.

17. This information stems from the species *Studbooks* and more particularly from the *International Studbook of the Orangutan (Pongo pygmaeus, Pongo abelii). 2010*, coordinated by Megan Elder, Como Zoo, Minnesota, and the European *Studbooks* of 2007 to 2010, *EEP für Orang-Utans. Europäisches Erhaltungszuchtprogramm. Zuchtbuch für Europa. XXV. 2007 & XXVII. 2010*, coordinated by Dr. Clemens Becker, Karlsruhe Zoo. The first directory of captive orangutans was published by the Yerkes Regional Primate Research Center in 1970.

18. Some information relating to capture of animals in their natural habitat remains vague. According to William Conway (director of the Bronx Zoo between 1962 and 1999 and president of the Wildlife Conservation Society, 1992–99), approximately 93% of captive mammals were born in zoos in the 2000s. There was, however, also an illegal market for protected species, under the control of specialized dealers and unscrupulous zoos. See Koen Margodt, *The Welfare Ark: Suggestions for a Renewed Policy in Zoos* (Brussels: VUB University Press, 2000).

19. Louis Montane, "A Cuban Chimpanzee," *Journal of Animal Behavior* 6 (1916): 330–33.

20. Irwin Bernstein, "Response to Nesting Materials of Wildborn and Captive Born Chimpanzees," *Animal Behaviour* 10 (1962): 1–6; Irwin Bernstein, "Age and Experience in Chimpanzees Nest Building," *Psychological Reports* 20 (1967): 1106; Irwin Bernstein, "A Comparison of Nesting Patterns among the Three Great Apes," in *The Chimpanzee*, Vol. 1, *Anatomy, Behavior, and Diseases of Chimpanzees*, ed. Geoffrey H. Bourne (Basel: Karger, 1969), 393–402.

21. Michel Serres, *Le Mal propre: Polluer pour s'approprier?* (Paris: Le Pommier, 2008), 17.

22. When they are injured, great apes construct nests that are higher up, in order to protect themselves from predators, but avoid risking a fall.

23. Bernard Richard, "Les mammifères constructeurs," in *La Recherche en éthologie. Les comportements animaux et humains*, ed. Jean-Pierre Desportes and Assomption Vloebergh (Paris: Éditions du Seuil, 1979), 156–73.

24. See the chapter "A as in Animal" in the documentary *L'Abécédaire de Gilles Deleuze* (interviewed by Claire Parnet), filmed by Pierre-André Boutang, 1996.

25. Gilles Deleuze and Félix Guattari, *A Thousand Plateaus: Capitalism and Schizophrenia* (London: Continuum, 1988).

26. Deleuze and Guattari, *Thousand Plateaus*, 311.

27. It is therefore important to exercise great caution when making decisions concerning them and to avoid making decisions in their stead without consulting them, in order to dictate "what is good for them," in a decidedly ideological manner. We are in fact sometimes subject to strange prejudices concerning those we rub shoulders with. We tend to speculate rather than examine them directly, and are then surprised when an unexpected event crops up and contradicts everything we had imagined to be true.

28. There are of course cases where the apes leave their cages and escape when they get an opportunity to do so. Nevertheless, the examples provided allow discussion of certain aspects that would otherwise remain obscure.

29. "What I'm trying to pick out with this term is, firstly, a thoroughly heterogeneous ensemble consisting of discourses, institutions, architectural forms, regulatory decisions, laws, administrative measures, scientific statements, philosophical, moral and philanthropic propositions–in short, the said as much as the unsaid. Such are the elements of the apparatus. The apparatus itself is the system of relations that can be established between these elements" (from the "The Confession of the Flesh" interview, 1977). In Michel Foucault, *Power/Knowledge Selected Interviews and Other Writings, 1972–1977*, ed. Colin Gordon (New York: Pantheon, 1980), 194–228.

30. Serres, *Mal Propre*, 18.

31. The psychologist Anne Russon (York University, Toronto), a specialist for this species, affirms that orangutans have a well-developed sense of *order*. They appreciate it when their environment is well organized.

32. For discussion of the relationships between trash, pollution, and territories, see Serres, *Mal propre*, 9–14.

33. "This is my home" in the sense that, as soon as a space is clearly defined, the occupant declares it as its *property*, an action which for Rousseau constitutes the foundation of civil society.

34. Deleuze and Guattari, *Thousand Plateaus*, 311.

35. This is not the case for all species or for all individuals. Some can be stressed to

the extreme by the presence of humans, neglecting their social life and showing greater aggression. Withdrawal zones should hence be provided within each enclosure. See Arnold S. Chamove, Geoffrey R. Hosey, and Peter Schaetzel, "Visitors Excite Primates in Zoos," *Zoo Biology* 7 (1988): 359–69.

36. Margaret Power, *The Egalitarians—Human and Chimpanzee: An Anthropological View of Social Organization* (Cambridge: Cambridge University Press, 1991).

37. Jonathan Knight, "The Naked Chef," *New Scientist* 2252 (2000): 6.

38. Samuel Fernandez-Carriba, "Los chimpancés que trituran la comida: un ejemplo de transformación del alimento en primates no humanos," *Boletín de la Asociación Primatológica Española* 7 (2000): 7–8; and Samuel Fernandez-Carriba and Angela Loeches, "Fruit Smearing by Captive Chimpanzees: A Newly Observed Food Processing Behavior," *Current Anthropology* 42 (2000): 143–147.

39. http://www.youtube.com/watch?v=NJngWS8iVRA.

40. *Exaptation* describes the evolution of an adaptive trait allowing a particular function to be served that emerged not because of natural selection linked to current function, but to initial selection in a different direction. Thus, the whole set of traits that would have favored moving around in a vertical position in trees, present in the primate ancestors of ancient hominids, were co-opted to fulfill a new function: bipedal locomotion. Traits that were selected for within the framework of an arboreal lifestyle hence allowed the emergence of bipedalism. Another example: bird feathers are thought to be an adaptive trait allowing birds to be protected and insulated so as to resist certain temperatures. These were later adapted for use in flight. See Stephen J. Gould and Elisabeth S. Vrba, "Exaptation—A Missing Term in the Science of Form," *Paleobiology* 8 (1982): 4–15.

41. This adjective is coined from the word "conciliance" cf. *A Latin-French Dictionary Félix Gaffiot*: *Dictionnaire illustré Latin-Français* (Paris: Hachette, 1934), 371: Conciliatio, onis, f.: association, union, harmony, fact of gaining the favor, fact of gaining the benevolence. Natural inclination, penchant. Note that the concept of *conciliance* is different than the concept of *consilience*, in particular developed by Edward O. Wilson, but the two notions (conciliance and consilience) are not unrelated.

42. Margodt, *Welfare Ark*, 86–87 (see "The Feasibility of Reintroductions").

43. David Hulme and Marshall W. Murphree, eds., *African Wildlife and Livelihoods: The Promise and Performance of Community Conservation* (Portsmouth, NH: Heinemann, 2001).

44. Of course, one must distinguish what is there to please the spectator from what really makes a difference for the animal. An appealing "presentation" will often mislead the visitor: it should above all be the animals that are consulted as the "experts" concerning evaluations of their habitats.

45. The San Diego Zoo Safari Park (California) uses a slightly different approach. As mentioned previously, the park has devoted several hectares to its "African savanna," which can be seen in its entirety only as part of a touristic train ride. The environment is that of Southern California, on the edge of the desert. Because of its vast dimensions, the space available to the animals here has little in common with the areas other zoos provide for their animals. Nevertheless, even in this very particular case, one must recognize that animal parks remain *artificial*: they represent enclaves that are elaborated, protected, and controlled by humans. We are not really in Africa, after all.

46. With increased interest when the protagonists in the films are representatives of their own species. In the USA, most zoos let their great apes view television or videos.

47. *World* in the sense of *Umwelt*, as proposed by Jakob von Uexküll (1909). See Jacob von Uexküll, *Mondes animaux et monde humain* (Paris: Médiations Denoël, 1984).

48. *Habitus,* which simultaneously encompasses the idea of an *acquired predisposition* and of *social imprinting* that together mold a personality (Norbert Elias); and of a *inclusive network,* which involves different dimensions: physical, psychological, social, and cultural (Marcel Mauss).

49. B. Van Puijenbroeck, former curator responsible for mammals at Antwerp Zoo (Royal Zoological Society of Antwerp, Belgium), accordingly maintains that animals kept in zoos are *impoverished* due to the *depleted nature* of their habitat, either social or environmental. For example, they no longer have to worry that a particular branch will break when they are using it while moving about; do not need to use complex mental representations of the zone in which they live; do not need to correctly memorize the positions of food sources in relation to seasonal fruiting periods. Personal communication, Antwerp, 2003.

50. In fact, in great apes, only very few of the behaviors that are necessary for survival in the wild seem to be inherited: Hilary Box in Margodt, *Welfare Ark,* 86–87.

51. Neotenic: retaining juvenile traits as an adult.

52. Hornaday, *Minds and Manners of Wild Animals,* 1.

53. Dolphins have also shown evidence for an attraction to iPads, as well as an ability to use them. This is also true for cats, one of the most famous being Choupette Lagerfeld, the one-year-old female Siamese belonging to Karl Lagerfeld.

54. See http://www.wptv.com/dpp/news/science_tech/apps-for-apes-program-comes-to-center-for-the-great-apesin-florida.

55. Peanut and Pumpkin are part of the very small circle of twin births of great apes to be born in zoos.

56. See Renato Bender and Nicole Bender, "Brief Communication: Swimming and Diving Behavior in Apes (*Pan troglodytes* and *Pongo pygmaeus*): First Documented Report," *American Journal of Physical Anthropology* 152 (2013): 156–62. On YouTube, see "Cooper the Chimp Learns to Swim" (six videos) and https://www.youtube.com/watch?v=o5i1xhq0G2Y

57. Some orangutans on Kaja Island (near Wanariset, the second rehabilitation station on Borneo) learned to swim. They were thus able to cross the river in order to eat fruits growing on the other bank.

58. Jane Goodall in *Visions of Caliban: On Chimpanzees and People,* ed. Dale Peterson and Jane Goodall (Boston: Houghton Mifflin, 1993), 292–93.

59. For example, approximately 184 types of plant have been identified as constituents of the natural diet of chimpanzees. These fruits are supplemented with insects, birds, eggs, and small mammals (Goodall, *Chimpanzees of Gombe*).

60. See the chapter "Beauty, the Bears, and the Setting Sun," in Jeffrey M. Masson and Susan McCarthy, *When Elephants Weep: The Emotional Lives of Animals* (New York: Delta, 1996), 192–211.

61. Peterson and Goodall, *Visions of Caliban,* 291–92.

62. See Lestel, *Amis de mes amis.*

Chapter Four

1. Heini Hediger, "Un problème qui nous ramène à l'homme: l'habitat des animaux," in Jacques Graven, *La pensée non humaine* (Paris: Planète, 1963).

2. William C. McGrew and Linda F. Marchant, "Chimpanzee Wears a Knotted Skin 'Necklace,'" *Pan Africa News* 5 (1998): 8–9 (newsletter published by Department of Zoology, Kyoto University).

3. Robert M. Yerkes and Ada Watterson-Yerkes, *The Great Apes: A Study of Anthropoid Life* (New Haven: Yale University Press, 1929).

4. Yerkes and Watterson-Yerkes, *The Great Apes*, 177.

5. Yerkes and Watterson-Yerkes, *The Great Apes*, 183.

6. Yerkes and Watterson-Yerkes, *The Great Apes*, 346.

7. Jacques Vauclair, "Would Humans without Language Be Apes?" in *Cultural Guidance in the Development of the Human Mind*, Vol. 7, *Advances in Child Development within Culturally Structured Environments*, ed. J. Valsiner and Aaro Toomela (Greenwich, CT: Ablex Publishing Corporation, 2003), 10–11.

8. I thank Cécile Chaussey for drawing my attention to Wattana's knot-making behaviors.

9. Various films have been devoted to showing Wattana's knot tying, including a research video consisting of 75 sequences of knotting (length: 1 hour 23 minutes): F. Gaillard and C. Herzfeld, *Singes noueurs. Le cas de Wattana, orang-outan* (Paris, 2008). See also C. Herzfeld, *Les noeuds de Wattana* (Exposition "Les grands singes vont-ils disparaître?"—Montage: Violette Araujo), Cité des sciences et de l'industrie (Paris, 2004); *Wattana, le singe qui sait faire des noeuds* (Grande Halle de la Villette), Exhibition "Bêtes et Hommes," (Paris, September 12, 2007–January 20, 2008); F. Gaillard and C. Herzfeld, *Funktionslust. Les noeuds de Wattana, orang-outan* (Paris, 2008) (two versions: *Technicité* 15 minutes and *Virtuosité* 9 minutes).

10. The length of these strands is extremely important. If they are too short, they offer only a few possibilities for performing knot tying. The strands that I gave to Wattana were about 5 feet long on average. However, strands of this length are also associated with a greater risk of strangulation.

11. See the first official publications on the topic of knot-making in great apes. Chris Herzfeld and Dominique Lestel, "Knot Tying in Great Apes: Etho-ethnology of an Unusual Tool Behavior," *Social Science Information* 44 (2005): 621–53; Dominique Lestel and Chris Herzfeld, "Topological Ape: Knots-Tying and Untying and the Origins of Mathematics," in *Images and Reasoning, Interdisciplinary Conference Series on Reasoning Studies*, Vol. 1, ed. Pierre Grialou, Giuseppe Longo, and Mitsuhiro Okada (Tokyo: Keio University Press, 2006), 147–62.

12. To produce such a knot, one must first shape a loop by crossing the two ends of a cord, then thread one of these ends through the loop. After that, it is only necessary to pull.

13. I thank Jacques Vauclair for drawing my attention to this aspect—personal communication, Paris, February 2004. See also François Bresson, "Inferences from Animal to Man: Identifying Functions," in *Methods of Inference from Animal to Human Behaviour*, ed. Mario von Cranach (Paris: Mouthon, 1976), 319–42.

14. Tim Ingold, *The Perception of the Environment: Essays on Livelihood, Dwelling and Skill* (London: Routledge, 2000). See chapters 18 and 19. (Chapter 18, "On Weaving a Basket," derives from an older article published in 1996 with the title "Making Culture and Weaving the World.")

15. Damien De Callataÿ, *Le Pouvoir de la gratuité: l'échange, le don, la grâce* (Paris: L'Harmattan, 2011), 166.

16. This study was performed in October, November, and December 2003, and allowed me to collect roughly a hundred meaningful responses from primatologists, instructors of talking apes, and animal technicians.

17. Biruté M. F. Galdikas, *Reflections of Eden: My Years with the Orangutans of Borneo* (New York: Little, Brown, 1995), 60–61.

18. The animal technician who looked after Wattana at Stuttgart Zoo, Margot

Federer, confirmed that Wattana did not yet know how to tie knots when she was in Germany, up to the time she left at the age of two-and-a-half.

19. Eric Baratay and Elisabeth Hardouin-Fugier, *Zoo: A History of Zoological Gardens in the West* (London: Reaktion Books, 2004), 79.

20. Andrew Whiten, personal communication, October 2003.

21. Anne E. Russon, *Orangutans: Wizards of the Rain Forest* (Toronto: Key Porter, 1999), 59.

22. Russon, *Orangutans*, 75.

23. Anne E. Russon and Birutė M. Galdikas, "Imitation in free-ranging rehabilitant orangutans (*Pongo pygmaeus*)." *Journal of Comparative Psychology* 107 (1993): 147–61.

24. Birutė M Galdikas, "Orang-utan Tool Use at Tanjung Putting Reserve, Central Sudonesian Borneo (Kalimantan Tengah)," *Journal of Human Evolution* 10 (1982): 19–33.

25. See Lyn H. Miles, Robert W. Mitchell, and Stephen E. Harper, "Simon Says: The Development of Imitation in an Enculturated Orangutan," in *Reaching into Thought: The Minds of the Great Apes*, ed. Anne E. Russon, Kim A. Bard, and Sue Taylor Parker (Cambridge: Cambridge University Press, 1996), 278–99.

26. Concerning learning through *peering*, see page 36.

27. Concerning the subject of "charismatic leaders" (Power, *Egalitarians*), see page 70.

28. *Funktion*: function, activity; *Lust*: joy, pleasure, desire, lust, inclination, urge.

29. Marc Richir, *Phénoménologie et institution symbolique* (Paris: Editions J. Millon, 1988), 264–65.

30. Jeffrey M. Masson and Susan McCarthy also mention this aspect—Masson and McCarthy, *When Elephants Weep*.

31. Étienne Souriau, *Le sens artistique des animaux* (Paris: Hachette, 1965).

32. Ibid.

33. Ibid.

34. Jürgen Lethmate, "Instrumental Behaviour of Zoo Orang-utans." *Journal of Human Evolution* 8 (1979): 741–44.

35. Several authors describe the same kind of aptitude: Hornaday (1922), Gewalt (1975), and Rumbaugh and Gill (1973).

36. See Jürgen Lethmate, "Instrumental Behavior," 744; Fritz Jantschke, *Orang-Utans in Zoologischen Gärten* (Munich: Piper, 1972); Duane M. Rumbaugh and Timothy V. Gill, "The Learning Skills of Great Apes," *Journal of Human Evolution* 2 (1973): 171–72. This claim is also at the core of arguments by Kortlandt (1968), who links primate behaviors to their ecological context—Adriaan Kortlandt, "Handgebrauch bei freilebenden Schimpansen," in *Handgebrauch und Verständigung bei Affen und Frühmenschen*, ed. Bernhard Rensch, 59–102 (Bern/Stuttgart: Huber, 1968).

37. Jürgen Lethmate, "Tool-using Skills of Orang-utans," *Journal of Human Evolution* 11 (1982): 49–64.

38. Ibid., 58.

39. Ibid., 57.

40. Just as for ancient hominids that also undoubtedly built "nests," ground nests presumably being at the origin of construction of huts and various other refuges. These constructions give rise to various questions, such as those concerning transport of materials, fastening elements, details of the structure, building techniques, tools for cutting, and choice of components. See Nold Egenter, "Evolutionary Architecture: The Nest Building Behavior of Higher Apes," *International Semiotic Spectrum* 14 (1990). http://home.worldcom.ch/~negenter/081NestbApes_E.html.

41. Most orangutans never descend to the ground.

42. Nests are only used once by great apes. However, they are sometimes re-used by other species, such as bushbabies.

43. Andrew Whiten, Jane Goodall, William C. McGrew, et al., "Cultures in Chimpanzees," *Nature* 399 (1999): 682–85; Carel P. Van Schaik, Marc Ancrenaz, Gwendolyn Borgen, et al., "Orangutan Cultures and the Evolution of Material Culture," *Science* 299 (2003): 102–5.

44. Dominique Lestel, *Les origines animales de la culture* (Paris: Flammarion, 2001), 280.

45. Lestel, *Origines animales de la culture,* 281.

46. Cf. Jean-Jacques Rousseau: "Where does one gain pleasure in such a situation? From nothing exterior to oneself, from nothing other than oneself and one's own existence." Jean-Jacques Rousseau, *Les rêveries du promeneur solitaire* (Genève, 1782).

47. *Axes de la Perfection. Les outils de pierre et la situation difficile du progrès, en archéologie de la Préhistoire, au XIXe siècle* (unpublished).

48. "We know, from all of the cultures that preceded *Homo sapiens,* from the stone tools that are practically our only evidence, that tools, overall, have followed a progressive evolutionary trajectory comparable to that followed by human anatomy, starting with the distant australopithecines, and passing through *Homo erectus* and the Neanderthals. . . . Tools are arranged along the timescale in an order that generally appears to be, both logical and chronological." André Leroi-Gourhan, *Évolution et Technique* (Paris: Albin Michel, 1943), 24.

49. Claudia Jordan, "Object Manipulation and Tool Use in Captive Pygmy Chimpanzees (*Pan paniscus*)," *Journal of Human Evolution* 11 (1982): 35–39.

50. Linda Van Elsacker and Vera Walraven, "The Spontaneous Use of a Pineapple as a Recipient by a Captive Bonobo (*Pan paniscus*)," *Mammalia* 58 (1994): 159–62.

51. Wolfgang W. Köhler, *The Mentality of Apes* (London: Kegan Paul, Trench, Trubner, 1925), 324.

52. Henry C. Raven, "Further Adventures of Meshie. A Chimpanzee That Has Lived Most of Her Life in a New York Suburban Home," *Natural History* 33 (1933): 607. (Reprinted: *Natural History,* July/August 2002).

53. It is possible that Wattana was also inspired by Nénette to learn how to tie knots, but she mainly observed humans, as her conspecific companion performed knotting only occasionally.

54. Lisa H. Abra, personal communication, November 2003.

55. Hilde Vervaecke, personal communication, November 2003.

56. Lynn Killam, personal communication, November 2003.

57. The female orangutan Deedee was born in Dallas on January 8, 1980, and was hand reared. She was later transferred to the zoo in Brownsville, also in Texas. She now lives in Lowry Park Zoo, in Tampa, Florida.

58. Lorin Milk, personal communication, 2003.

59. Personal observation, Lowry Park Zoo, Tampa, Florida, August 30, 2003.

60. Nannette Driver, personal communication, November 2003.

61. Marina Vancatova, personal communication, August 31, 1999.

62. Moja was born on November 18, 1972. She died in 2002.

63. Deborah Fouts, personal communication, December 2003.

64. Lyn Miles, personal communication, Atlanta, May 2004.

65. Personal observation, Atlanta Zoo, May 2004.

66. E. Linden, *The Parrot's Lament.*

67. Michael Beran—contacted via David Washburn, personal communication, December 2003.

68. Personal observation, Language Research Center, Atlanta, 2003.

69. Bill Fields, personal communication, November 2003.

70. Francine Patterson—contacted via Sandra Marchese, personal communication, December 2003.

71. Or even siamangs, *Symphalangus syndactylus* (Orgeldinger, 1999) and talapoins, *Miopithecus talapoin* (Zimmerman, personal communication, October 2003).

72. We have not counted Koko, the only gorilla on the list, as the observations we had access to were not sufficiently conclusive.

73. Russon, *Orangutans*, 46.

74. Philippe Muller, in von Uexküll, *Mondes animaux et monde humain*, 7.

75. As regards the importance of the body, see work by Jocelyne Porcher. For example, Porcher, "Le travail dans l'élevage industriel des porcs. Souffrance des animaux, souffrance des hommes," in *Les animaux d'élevage ont-ils droit au bien-être?*, ed. Florence Burgat and Robert Dantzer (Paris: INRA, 2001), 59–60.

76. Konrad Lorenz, *King Solomon's Ring* (London: Methuen, 1952) (1949 for the original German version).

Chapter Five

1. According to Egenter, the German words "Werk" and "wirken," the English "work," and the Greek "Ergon" are derived from a group of terms meaning to turn, fold, roll up, and weave.

2. Cf. Latin: *artis facta. Ars*, artis, *f.* (1) talent, know-how, skill—(2) anything to which the talent or skill is applied—(3) technical knowledge, theories, bodies of doctrine or art // *Factum*, i, m. action, enterprise, work.

3. In the sense that attention is paid to the aesthetics of daily life and to certain everyday objects (like nests) and not in the sense of being linked to the artistic *movement* initiated by William Morris in the mid-nineteenth century.

4. Desmond Morris, *The Naked Ape* (New York: McGraw-Hill, 1968).

5. Desmond Morris, *The Biology of Art. A Study of the Picture-making Behaviour of the Great Apes and Its Relationship to Human Art* (London: Methuen, 1962).

6. Thierry Lenain, *Monkey Painting* (London: Reaktion Books, 1997), 80. Translated by Caroline Beamish.

7. Edward O. Wilson, *Sociobiology* (Cambridge: Harvard University Press, 1980).

8. Lenain, *Monkey Painting*, 90.

9. Desmond Morris cited by Lenain, *Monkey Painting*, 91.

10. Ibid.

11. Lenain, *Monkey Painting*, 91.

12. Lilo Hess, *Christine, the Baby Chimp* (London: Bell, 1954).

13. Lenain, *Monkey Painting*, 69.

14. Lenain, *Monkey Painting*, 76.

15. Lenain, *Monkey Painting*, 165.

16. Vitaly Komar and Aleksandr Melamid, *When Elephants Paint: The Quest of Two Russian Artists to Save the Elephants of Thailand* (New York: Harper, 2000).

17. See http://www.elephantart.com/catalog/. This type of commercial operation is also found in India and on Bali.

18. David Gucwa and James Ehmann, *To Whom It May Concern: An Investigation of the Art of Elephants* (New York: Norton, 1985).

19. Siri was captured in Thailand when about two years old.

20. Gucwa and Ehmann, *To Whom It May Concern*, 120.

21. Souriau, *Sens artistique des animaux*, 26.

22. Olivier Messiaen recorded and transcribed the songs of some species that he described not only as virtuosos, but also as artists. He then integrated these into some of his compositions, notably in *Réveil des oiseaux* (1953), *Oiseaux exotiques* (1956), *Catalogues d'oiseaux* (1956–58), and *Chronochromie* (1960).

23. Deleuze and Guattari, *Thousand Plateaus*, 316–17.

24. See François-Bernard Mâche, *Musique au singulier* (Paris: Odile Jacob, 2001); François-Bernard Mâche, *Music, Myth, and Nature, or, the Dolphins of Arion* (Chur, Switzerland: Harwood, 1992).

25. George Herzog, "Do Animals Have Music?" *Bulletin of the American Musicological Society* 5 (1941): 3–4.

26. There are some exceptions among scientists, for example, O. Koelher, W. H. Thorpe, J. Hall-Craggs, or O. Sotavalta, a zoologist with absolute pitch. See particularly Joan Hall-Craggs, "The Aesthetic Content of Bird Song," in *Bird Vocalizations*, ed. Robert A. Hinde (Cambridge: Cambridge University Press, 1969), 367–381.

27. François-Bernard Mâche, "Les oiseaux musiciens," *Sciences et Avenir, Hors-Série Paroles Animales*, 131 (2002): 62–68.

28. Rosemary Jellis, *Bird Sounds and their Meaning* (Ithaca, NY: Cornell University Press, 1977).

29. Charles Hartshorne, *Born to Sing: An Interpretation and World Survey of Bird Song* (Bloomington: Indiana University Press, 1973).

30. Peter Szöke, "Ornitomuzikologia," *Magyar Tudomany* 9 (1963): 592–607.

31. Souriau, *Sens artistique des animaux*, 62.

32. Mâche, *Musique au singulier*, 284.

33. Mâche, "Les oiseaux musiciens," 67.

34. Mâche, *Musique au singulier*, 284.

35. Souriau, *Sens artistique des animaux*, 61–62.

36. See Hollis Taylor's website: http://www.zoomusicology.com. Notably, the site lists a bibliography that is relevant to the topics discussed in this paragraph.

37. Christian Demars and Michel Goustard, "Structure et règles de déroulement des émissions sonores des Hylobates, *Hylobates concolor*," *Bulletin Biologique de la France et de la Belgique* 106 (1973): 177–91.

38. For example, see Dario Martinelli, "How Musical Is a Whale? Towards a Theory of Zoömusicology," in *Acta Semiotica Fennica: Approaches to Musical Semiotics*, edited by E. Tarasti, Hakapaino, International Semiotics Institute (Helsinki: Semiotic Society of Finland, 2002).

39. See his website: http://www.interspecies.com/pages/animalmusic.html

40. Alan Hovahaness (1994), *And God Created Great Whales, for Orchestra and Taped Whale Sounds*, Op. 229, No 1.

41. See Ken Marten, Karim Shariff, Suchi Psarakos, and Don White, "Les ronds d'air des dauphins," *Pour la Science* 228 (1996): 90–95.

42. Project Delphis, Earthtrust Association.

43. Some bubbles have a diameter of more than 2 feet.

44. Don White, "Mystery of the Silver Rings." See http://earthtrust.org/archive/delrings.html

45. William Jardine and Prideaux J. Selby, *Illustrations of Ornithology*, Vol. 1 (Edinburgh: D. Lizars, 1837).

46. Concerning bowerbirds, Étienne Souriau recommends the work of Alec H. Chisholm. See, for example, Alec H. Chisholm, *Observations on the Golden Bower-Bird* (Sydney: Royal Australian Ornithologist Union, 1956).

47. John Gould and Elizabeth Gould, *Birds of Australia* (London, 1840–1848). Available online: http://www.nla.gov.au/apps/cdview?pi=nla.aus-f4773.

48. Golden-fronted bowerbird, streaked bowerbird, white-eared catbird, great bowerbird, fawn-breasted bowerbird, Vogelkop bowerbird, Archbold's bowerbird, fire-maned bowerbird, yellow-breasted bowerbird, MacGregor's bowerbird, golden bowerbird, Sanford's bowerbird, flame bowerbird, spotted bowerbird, spotted catbird, regent bowerbird, satin bowerbird, western bowerbird, green catbird, and, finally, the stagemaker bowerbird (the species mentioned by Deleuze and Guattari in *A Thousand Plateaus*). The term "bowerbird" is sometimes replaced with "catbird."

49. Although bowerbirds sometimes live very far from the rivers where one find the shells they use in constructing their stages.

50. Reflection of the naturalist Odoardo Beccari, during a voyage to New Guinea accompanying the ornithologist Luigi Maria d'Albertis (1872).

51. This explanation applies to only some of the bowerbird species, as certain representatives are actually very colorful.

52. Thomas E. Gilliard, "Bower Ornamentation versus Plumage Characters in Bower-Birds," *Auk* 73, no. 3 (1956): 450–51.

53. James L. Gould and Carol Grant Gould, *Animal Architects: Building and the Evolution of Intelligence* (New York: Perseus, 2007).

54. Joah R. Madden, Caroline Dingle, Jess Isden, et al., "Male Spotted Bowerbirds Propagate Fruit for Use in Sexual Display," *Current Biology* 22 (2012): R264–65.

55. See the chapter "Beauty, the Bears, and the Setting Sun" in Jeffrey M. Masson and Susan McCarthy, *When Elephants Weep: The Emotional Lives of Animals* (New York: Delta, 1996), 192–211.

56. Souriau distinguishes two types of aesthetic sensitivity: the *emitting sensitivity* of the creator and the *receptive sensitivity* of the one that looks at the work. These two types of sensitivity "collaborate": "The sensitivity of the emitter frequently needs control of the sensitivity of the receiver"—Souriau, *Sens artistique des animaux*, 51.

57. Michael H. Hansell, *Built by Animals: The Natural History of Animal Architecture* (New York: Oxford University Press, 2007), chapter 8, "Beautiful Bowers," p. 235.

58. John A. Endler, Lorna C. Endler, and Natalie R. Doerr, "Great Bowerbirds Create Theaters with Forced Perspective When Seen by Their Audience," *Current Biology* 20 (2010): 1679–84.

59. Laura Kelley and John Endler, "Illusions Promote Mating Success in Great Bowerbirds," *Science* 335 (2012): 335–38.

60. Brian Massumi, "Ceci n'est pas une morsure: Animalité et abstraction chez Deleuze et Guattari," *Philosophie* 112 (2011), 80.

61. Deleuze and Guattari, *Thousand Plateaus*, chapter "1837: Of the Refrain," 310–50.

62. Deleuze and Guattari, *Thousand Plateaus*, 315.

63. Deleuze and Guattari, *Thousand Plateaus*, 331.

64. Deleuze and Guattari, *Thousand Plateaus*, 315.

65. Karl Von Frisch and Otto Von Frisch, *Animal Architecture* (New York: Harcourt Brace Jovanovich, 1974), 243–44.

66. I thank Nathan Herzfeld for drawing my attention to numerous points raised in this section.

67. See Masson and McCarthy, *When Elephants Weep*, 145.

68. Jaak Panksepp, "Affective Consciousness: Core Emotional Feelings in Animal and Humans," *Consciousness and Cognition* 14 (2005): 30–80.

69. Jaak Panksepp, "On the Embodied Neural Nature of Core Emotional Affects," *Journal of Consciousness Studies* 12 (2005): 158–84.

70. Souriau, *Sens artistique des animaux*, 30.

71. Ibid.

72. Panksepp, "Affective Consciousness."

73. John E. Lisman and Anthony A. Grace, "The Hippocampal-VTA Loop Controlling Entry into Long-Term Memory," *Neuron* 46 (2005): 703–13; Amy Maxman, "Dopamine's Role Linked to Emotions," *Science News*, August 2, 2008.

74. Temple Grandin and Catherine Johnson, *Animals Make Us Human* (Boston: Houghton Mifflin Harcourt, 2009).

75. James Olds, *Drives and Reinforcements: Behavioral Studies of Hypothalamic Functions* (New York: Raven Press, 1977).

76. Souriau, *Sens artistique des animaux*, 30.

77. Deleuze and Guattari, *Thousand Plateaus*, 316.

78. Massumi, "Ceci n'est pas une morsure," 68.

79. Massumi, "Ceci n'est pas une morsure," 67.

80. Massumi, "Ceci n'est pas une morsure," 69.

81. Massumi, "Ceci n'est pas une morsure," 84.

82. Hansell, *Built by Animals*, 235.

83. Moreover, social scientists generally refuse to assign to animals those competences that they think are exclusively linked to culture, and hence privileges of humans.

84. Exploring and becoming affiliated with life constitutes a deep and complex process in mental development. For Wilson, *biophilia* is the "innate tendency to focus on life and lifelike processes." Edward Wilson, *Biophilia* (Cambridge, MA: Harvard University Press, 1984), 10.

85. Pierre Sterckx, *Le Devenir-cochon de Wim Delvoye* (Bruxelles: La Lettre volée, 2007).

86. What better example of this kind of art, bearer of *life itself*, than the works of Pollock, who declares: "I am nature" or "My concern is with the rhythms of nature"? Fabrice Midal, *Jackson Pollock ou L'invention de l'Amérique* (Paris: Éditions du Grand Est, 2008), 41. Painting as if he were dancing, totally immersed in this act of painting, Pollock indeed lets himself be swept up by the flows of life: he does not *represent* this flux, but his creative gestures, his *drippings*, his traces, his marks *are* the flux itself. During the immediacy of creation, which becomes choreography and meditation, he attempts to connect himself to the universal vital forces, physically and spiritually, and projects these onto the canvas.

87. Catalogue of the "Jackson Pollock et le Chamanisme" exhibition, Pinacothèque de Paris, Paris, 15.10.2008–15.02.2009, 11.

88. See the conclusions of Lenain, *Monkey Painting*.

89. Souriau, *Sens artistique des animaux*, 8.

90. Deleuze and Guattari, *Thousand Plateaus*, 320.

91. Deleuze and Guattari, *Thousand Plateaus*, 320.

92. Souriau, *Sens artistique des animaux*, 8.

93. This uniqueness is notably manifested in the extraordinary diversity of artistic practices in humans, by the development in techniques for some art forms, through their

inscription in a history, by the reflexivity of the artist, by a particular connection with the sacred, by the manifestation of a conscience that is amplified by the use of language.

94. Aesthetics of daily life, which Souriau distinguishes from aesthetics centering on masterpieces and from a definition of art that would include only the most sublime works. Souriau, *Sens artistique des animaux*, 44–45.

95. Souriau, *Sens artistique des animaux*, 57.

Epilogue

1. See the conclusions in Chris Herzfeld and Dominique Lestel, "Knot Tying in Great Apes," *Social Science Information* 44 (2005): 621–53.

2. Quote is from Donna L. Hart and Robert W. Sussman, *Man the Hunted: Primates, Predators, and Human Evolution* (Cambridge, MA: Westview, 2005).

3. Amos was born on December 20, 2000, in Usti (Czech Republic). He arrived in Apeldoorn on July 24, 2008.

4. They were taken away from their mother and hand-reared: http://www.youtube.com/watch?v=7gqvJgLDxv8

5. This was notably the case of the female Bonnie, at Washington Zoo (Washington, DC). She had learned that her plastic toys could be used to short-circuit the electrical system.

6. Étienne Souriau, *Le sens artistique des animaux* (Paris: Hachette, 1965), 106.

BIBLIOGRAPHY

Books

Baratay, Eric, and Elisabeth Hardouin-Fugier. *Zoo: A History of Zoological Gardens in the West.* London: Reaktion Books, 2004.

Bernardin de Saint-Pierre, Jacques-Henri. *Mémoire sur la nécessité de joindre une ménagerie au Jardin national des Plantes de Paris.* [*Memoir on the Need to Include a Menagerie in the Jardin National des Plantes in Paris.*] Paris: De l'imprimerie de Didot le Jeune, 1792.

Blanckaert, Claude, Claudine Cohen, Pietro Corsi, and Jean-Louis Fischer, eds. *Le Muséum au premier siècle de son histoire.* [*The First Century in the History of the Museum.*] [actes du colloque de Paris, juin 1993, Centre Alexandre Koyré], éd. du MNHN, coll. "Archives." Paris: Éditions du Muséum national d'Histoire naturelle, 1997.

Boitard, Pierre. *Le Jardin des Plantes. Description et mœurs des mammifères de la Ménagerie et du Muséum d'Histoire naturelle* (Précédé d'une introduction historique, descriptive et pittoresque par m. J. Janin). [*The Jardin des Plantes: Description and Behavior of the Mammals in the Menagerie at the Natural History Museum.* (Preceded by a historical, descriptive and pictorial introduction by M.J. Janin.).] Paris: Gustave Barba, 1863.

Bomsel, M.-C., J.-L. Berthier, S. Peron, and E. Goix. *La Ménagerie du Jardin des Plantes de Paris.* Guide-catalogue. [*The Menagerie in the Paris Jardin des Plantes: Guide and Catalog.*] Paris: Éditions du Muséum national d'Histoire naturelle, 2001.

Catherinet de Villemarest, Charles-Maxime. *Mémoires de Mademoiselle Avrillon: première femme de chambre de l'impératrice, sur la vie privée de Joséphine, sa famille et sa cour.* [*Memoirs of Miss Avrillon, the Empress's*

First Chambermaid: On the Private Life of Josephine, her Family and her Court.] Paris: Garnier frères, 1896.

Chisholm, Alec H. *Observations on the Golden Bower-Bird.* Sydney: Royal Australian Ornithologist Union, 1956.

Coupin, Henri. *Singes et singeries.* [*Monkeys and Monkey Business.*] Paris: Vuibert & Nony, 1907.

De Callataÿ, Damien. *Le Pouvoir de la gratuité: l'échange, le don, la grâce.* [*The Power of Free Access: Exchange, Giving and Grace.*] Paris: L'Harmattan, 2011.

De Fontenay, Elisabeth. *Le silence des bêtes. La philosophie à l'épreuve de l'animalité.* [*The Silence of Beasts: Philosophy Faced by Animality.*] Paris: Fayard, 1998.

Deleuze, Gilles. *Francis Bacon. The Logic of Sensation.* Translated by Daniel W. Smith. London/New York: Continuum, 2004.

Deleuze, Gilles, and Félix Guattari. *Kafka: Toward a Minor Literature.* Translated by Dana Polan. Minnesota: University of Minnesota Press, 1986.

Deleuze, Gilles, and Félix Guattari. *A Thousand Plateaus: Capitalism and Schizophrenia.* Translated by Brian Massumi. London: Continuum, 1987.

De Waal, Frans B. M. *Chimpanzee Politics: Power and Sex among Apes.* Baltimore: Johns Hopkins University Press, 1982.

Dutton, Denis. *The Art Instinct: Beauty, Pleasure, and Human Evolution.* New York: Bloomsbury, 2010.

Fossey, Dian. *Gorillas in the Mist.* Boston: Houghton Mifflin, 1983.

Fouts, Roger, and Stephen Tukel Mills. *Next of Kin: My Conversations with Chimpanzees.* New York: William Morrow, 1997.

Galdikas, Birute M. F. *Reflections of Eden: My Years with the Orangutans of Borneo.* New York: Little, Brown, 1995.

Gay, Pierre. *Des zoos pour quoi faire? Pour une nouvelle philosophie de la conservation.* [*What Are Zoos For? Toward a New Philosophy of Conservation.*] Paris: Delachaux & Niestlé, 2005.

Goodall, Jane. *The Chimpanzees of Gombe: Patterns of Behaviour.* Cambridge, MA: Belknap Press, 1986.

Gould, James L., and Carol Grant Gould. *Animal Architects: Building and the Evolution of Intelligence.* New York: Perseus, 2007.

Gould, John, and Elizabeth Gould. *Birds of Australia.* London, 1840–1848.

Grandin, Temple, and Catherine Johnson. *Animals Make Us Human.* Boston: Houghton Mifflin Harcourt, 2009.

Graven, Jacques. *La pensée non humaine.* [*Nonhuman Thought.*] Encyclopédie Planète. Paris: Planète, 1963.

Gucwa, David, and James Ehmann. *To Whom It May Concern: An Investigation of the Art of Elephants.* New York: Norton, 1985.

Hansell, Michael H. *Animal Architecture and Building Behavior.* London: Longman, 1984.

Hansell, Michael H. *Built by Animals: The Natural History of Animal Architecture.* New York: Oxford University Press, 2007.

Haraway, Donna. *Modest Witness@Second Millenium. Femaleman Meets Oncomouse: Feminism and Technoscience.* New York: Routledge, 1997.

Hart, Donna L., and Robert W. Sussman. *Man the Hunted: Primates, Predators, and Human Evolution.* Cambridge: Westview, 2005.

Hartshorne, Charles. *Born to Sing: An Interpretation and World Survey of Bird Song.* Bloomington: Indiana University Press, 1973.

Herzfeld, Chris. *Petite histoire des grands singes.* [*A Short History of the Great Apes.*] Paris: Éditions du Seuil, 2012.

Hess, Lilo. *Christine, the Baby Chimp.* London: Bell, 1954.

Hornaday, William T. *The Minds and Manners of Wild Animals: A Book of Personal Observations.* New York: Scribner's, 1922.

Hornaday, William T. *Popular Official Guide to the New York Zoological Park.* 18th ed. New York: New York Zoological Society, 1923.

Hulme, David, and Marshall W. Murphree, eds. *African Wildlife and Livelihoods: The Promise and Performance of Community Conservation.* Portsmouth, NH: Heinemann, 2001.

Ingold, Tim. *The Perception of the Environment: Essays on Livelihood, Dwelling and Skill.* London: Routledge, 2000.

Jantschke, Fritz. *Orang-Utans in Zoologischen Gärten.* [*Orangutans in Zoological Gardens.*] Munich: Piper, 1972.

Jardine, William, and Prideaux J. Selby. *Illustrations of Ornithology.* Vol. 1. Edinburgh: D. Lizars, 1837.

Jauffret, Louis-François. *Voyage au Jardin des Plantes, contenant la description des galeries d'histoire naturelle, des serres où sont renfermés les arbrisseaux étrangers, de la partie du jardin appelée l'école de botanique; avec l'histoire des deux éléphans, et celle des autres animaux de la ménagerie nationale.* [*Voyage to the Jardin des Plantes, Containing a Description of the the Galleries of Natural History, of the Greenhouses Enclosing the Exotic Shrubs, of the Part of the Garden Known as the School of Botany; Together with an Account of the Two Elephants and of the Other Animals in the National Menagerie.*] Paris: De l'Imprimerie de Ch. Houel, 1797–1798.

Jellis, Rosemary. *Bird Sounds and Their Meaning.* Ithaca, NY: Cornell University Press, 1977.

Jouin, Henry, and Henri Stein. *Histoire et description du Jardin des Plantes et du Muséum d'Histoire naturelle.* [*History and Description of the Jardin des Plantes and the Natural History Museum.*] Paris: Librairie Plon, 1887.

Jullien, François. *Le pont des singes: De la diversité à venir. Fécondité culturelle face à identité nationale.* [*The Monkey Bridge: On Future Diversity—Cultural Fecundity Face-to-Face with National Identity.*] Paris: Galilée, 2010.

Jullien, François. *The Propensity of Things: Toward a History of Efficacy in China.* New York: Zone Books, 1995.

Jullien, François. *A Treatise on Efficacy: Between Western and Chinese Thinking.* Honolulu: University of Hawaii Press, 2004.

Kaplan, Gisela T., and Lesley J. Rogers. *The Orangutans: Their Evolution, Behavior, and Future.* Cambridge, MA: Perseus, 2000.

Köhler, Wolfgang W. *The Mentality of Apes.* London: Kegan Paul, Trench, Trubner, 1925.

Komar, Vitaly, and Aleksandr Melamid. *When Elephants Paint: The Quest of Two Russian Artists to Save the Elephants of Thailand.* New York: Harper, 2000.

Latour, Bruno. *Pandora's Hope: Essays on the Reality of Science Studies.* Cambridge, MA: Harvard University Press, 1999.

Le Bras-Chopard, Armelle. *Le zoo des philosophes: De la bestialisation à l'exclusion.* [*The Philosophers' Zoo: From Bestialization to Exclusion.*] Paris: Plon, 2000.

Lenain, Thierry. *Monkey Painting.* London: Reaktion Books, 1997.

Leroi-Gourhan, André. *Évolution et technique.* [*Evolution and Technique.*] Paris: Albin Michel, 1943.

Lestel, Dominique. *Les amis de mes amis.* [*Friends of My Friends.*] Paris: Éditions du Seuil, 2007.

Lestel, Dominique. *Les origines animales de la culture.* [*The Animal Origins of Culture.*] Paris: Flammarion, 2001.

Lestel, Dominique. *Paroles de singes: L'impossible dialogue homme-primate.* Textes à l'appui/Série Sciences cognitives. [*Words of Apes: The Impossibility of Human-Primate Dialogue.*] Paris: Éditions de la Découverte, 1995.

Lévêque, Jean-Charles. *Le sens du beau chez les bêtes.* [*The Sense of Beauty in Animals.*] Paris: Éditions Villarrose, 2000.

Linden, Eugene. *The Octopus and the Orangutan.* New York: Dutton, 2002.

Linden, Eugene. *The Parrot's Lament and Other Tales of Animal Intrigue, Intelligence, and Ingenuity.* New York: Plume, 2000.

Loisel, Gustave. *Projet de réorganisation de la Ménagerie du Muséum.* [*Reorganization Project for the Museum's Menagerie.*] Paris: Imprimerie générale Lahure, 1906.

Loisel, Gustave. *Histoire des ménageries, de l'Antiquité à nos jours.* [*The History of Menageries from Antiquity to the Present Day.*] 3 vol. Paris: Octave Doin et fils et Henri Laurens, 1912.

Lorenz, Konrad. *King Solomon's Ring.* London: Methuen, 1952.

Mâche, François-Bernard. *Music, Myth, and Nature, or, the Dolphins of Arion.* Chur, Switzerland: Harwood, 1992.

Mâche, François-Bernard. *Musique au singulier.* [*Music in the Singular.*] Paris: Odile Jacob, 2001.

Margodt, Koen. *The Welfare Ark: Suggestions for a Renewed Policy in Zoos.* Brussels: VUB University Press, 2000.

Masson, Jeffrey M., and Susan McCarthy. *When Elephants Weep: The Emotional Lives of Animals.* New York: Delta, 1996.

McGrew, William C., Linda F. Marchant, and Toshisada Nishida, eds. *Great Ape Societies.* Cambridge: Cambridge University Press, 1996.

Michelet, Jules. *The People.* Translated by G. H. Smith. London, 1846.

Midal, Fabrice. *Jackson Pollock ou l'invention de l'Amérique.* [*Jackson Pollock or the Invention of America.*] Paris: Éditions du Grand Est, 2008.

Millin, Aubin L., Philippe Pinel, and Alexandre Brogniart. *Rapport fait à la Société d'Histoire Naturelle de Paris sur la nécessité d'établir une ménagerie, Paris, 14 décembre 1792, l'an 1er de la République.* [*Report Made to the Société d'Histoire Naturelle in Paris: On the Need to Establish a Menagerie. Paris 14th December 1792, First Year of the Republic.*] Paris: Imprimerie de Boileau, 1792.

Morris, Desmond. *The Biology of Art: A Study of the Picture-Making Behaviour of the Great Apes and Its Relationship to Human Art.* London: Methuen, 1962.

Morris, Desmond. *The Naked Ape.* New York: McGraw-Hill, 1968.

Olds, James. *Drives and Reinforcements: Behavioral Studies of Hypothalamic Functions.* New York: Raven Press, 1977.

Orgeldinger, Mathias. *Paarbeziehung beim Siamang-Gibbon* (Hylobates syndactylus) *im Zoo: Untersuchungen über den Einfluß von Jungtieren auf die Paarbindung.* [*Pair Relationship of the Siamang* (Hylobates syndactylus) *in the Zoo: Investigations of the Influence of Young Animals on Pair Bonding.*] Münster: Schüling Verlag, 1999.

Panksepp, Jaak. *Affective Neuroscience: The Foundation of Human and Animal Emotions.* New York: Oxford University Press, 1998.

Peterson, Dale, and Jane Goodall. *Visions of Caliban: On Chimpanzees and People.* Boston: Houghton Mifflin, 1993.

Picq, Pascal, Dominique Lestel, Vinciane Despret, and Chris Herzfeld. *Les grands singes: L'humanité au fond des yeux.* [*The Great Apes: Humanity in the Depths of the Eyes.*] Paris: Éditions Odile Jacob, 2005.

Porcher, Jocelyne. *Eleveurs et animaux: Réinventer le lien.* [*Keepers and Animals: Reinventing the Bond.*] Paris: Presses universitaires de France, 2002.

Power, Margaret. *The Egalitarians—Human and Chimpanzee: An Anthropological View of Social Organization.* Cambridge: Cambridge University Press, 1991.

Rapoport, Amos. *House Form and Culture.* New York: Prentice-Hall, 1969.

Richir, Marc. *Phénoménologie et institution symbolique.* [*Phenomenology and Symbolic Institution.*] Paris: Editions J. Millon, 1988.

Rilke, Rainer M., and Balthus. *Mitsou, Histoire d'un chat.* [*Mitsou: The Story of a Cat.*] Paris: Payot & Rivages, 2008.

Rousseau, Jean-Jacques. *Les rêveries du promeneur solitaire.* [*The Daydreams of a Solitary Walker.*] Genève, 1782.

Russon, Anne E. *Orangutans: Wizards of the Rain Forest.* Toronto: Key Porter, 1999.

Russon, Anne E., Kim A. Bard, and Sue Taylor Parker. *Reaching into Thought: The Minds of the Great Apes.* Cambridge: Cambridge University Press, 1996.

Schuster, Gerd, Willie Smits, and Jay Ullal. *Thinkers of the Jungle: The Orangutan Report.* Königswinter: Ullmann/Tandem, 2008.

Serres, Michel. *Malfeasance: Appropriation through Pollution?* Translated by Anne-Marie Feenberg-Dibon. Stanford: Stanford University Press, 2010.

Souriau, Étienne. *Le sens artistique des animaux.* [*The Artistic Sense of Animals.*] Paris: Hachette, 1965.

Stengers, Isabelle. *The Invention of Modern Science.* Minneapolis: University of Minnesota Press, 2000.

Stengers, Isabelle, and Judith E. Schlanger. *Les concepts scientifiques: invention et pouvoir.* [*Scientific Concepts: Invention and Power.*] Collection Folio/essais. Paris: Gallimard, 1991.

Sterckx, Pierre. *Le devenir-cochon de Wim Delvoye.* [*Wim Delvoye Turns Nasty.*] Bruxelles: La Lettre volée, 2007.

Strum, Shirley C. *Almost Human: A Journey into the World of Baboons.* New York: Random House, 1987.

Strum, Shirley C., and Linda M. Fedigan. *Primate Encounters: Models of Science, Gender, and Society.* Chicago: University of Chicago Press, 2000.

Thorpe, William H. *Learning and Instinct in Animals.* London: Methuen, 1956.

Van Schaik, Carel. *Among Orangutans: Red Apes and the Rise of Human Culture.* Cambridge, MA: Belknap, 2004.

Von Frisch, Karl, and Otto von Frisch. *Animal Architecture.* New York: Harcourt Brace Jovanovich, 1974.

Von Uexküll, Jacob. *Mondes animaux et monde humain.* [*Animal Worlds and the Human World.*] Paris: Médiations Denoël, 1984.

Vosmaer, Arnout. *Description de l'Espèce de Singe aussi singulier que très rare, nommé Orang-Outan, de l'Isle de Bornéo. Apporté vivant dans la Ménagerie de Son Altesse Sérénissime, Monseigneur le Prince d'Orange et de Nassau, Stad- houder Héréditaire, Gouverneur, Capitaine Général et Amiral des Provinces-Unies des Pais-Bas, Etc. Etc., Etc.* [*Description of the Rare and Distinctive Ape Species Called Orang-Outan, from the Island of Borneo. Delivered Alive into la Menagerie of His Serene Highness, His Eminence the Prince of Orange and Nassau, Hereditary Governor, Captain General and Admiral of the United Provinces of the Netherlands, etc., etc., etc..*] Amsterdam: Chez Pierre Meijer, 1778.

Wilson, Edward O. *Biophilia*. Cambridge, MA: Harvard University Press, 1984.
Wilson, Edward O. *Sociobiology*. Cambridge, MA: Harvard University Press, 1980.
Wrangham, Richard W., William C. McGrew, Frans B. M. de Waal, and Paul G. Heltne, eds. *Chimpanzee Cultures*. Cambridge, MA: Harvard University Press, 1996.
Yerkes, Robert M., and Ada Watterson-Yerkes. *The Great Apes: A Study of Anthropoid Life*. New Haven: Yale University Press, 1929.
Zuckerman, Solly. *The Social Life of Monkeys and Apes*. London: Kegan Paul, 1932.

Articles and Chapters

Baratay, Eric. "Belles captives: une histoire des zoos du côté des bêtes." [Beautiful Captives: A History of Zoos Seen from an Animal Perspective.] In *Beauté animale* (catalogue de l'exposition) [*Animal Beauty* (Catalog of the Exhibition)], edited by Emmanuelle Héran, 196–209. Paris: Réunion des musées nationaux/Grand Palais, 2012.
Barsanti, Giulio. "L'orang-outan déclassé. (*Pongo wurmbii* Tied.). Histoire du premier singe à hauteur d'homme (1780–1801) et ébauche d'une théorie de la circularité des sources." [The Orangutan Downgraded: Story of The First Primate of Human Stature (1780–1801).] *Bulletins et Mémoires de la Société d'Anthropologie de Paris* 1 (1989): 67–104.
Bender, Renato, and Nicole Bender. "Brief Communication: Swimming and Diving Behavior in Apes (*Pan troglodytes* and *Pongo pygmaeus*): First Documented Report." *American Journal of Physical Anthropology* 152 (2013): 156–62.
Bernstein, Irwin S. "Age and Experience in Chimpanzees Nest Building." *Psychological Reports* 20 (1967): 1106.
Bernstein, Irwin S. "A Comparison of Nesting Patterns among the Three Great Apes." In *The Chimpanzee*. Vol. 1, *Anatomy, Behavior, and Diseases of Chimpanzees*, edited by Geoffrey H. Bourne, 393–402. Basel: Karger, 1969.
Bernstein, Irwin S. "Response to Nesting Materials of Wildborn and Captive Born Chimpanzees." *Animal Behaviour* 10 (1962): 1–6.
Boesch, Christophe. "Teaching among Wild Chimpanzees." *Animal Behaviour* 41 (1991): 530–32.
Bourdelle, Edouard, and Paul Rode. "Note à propos d'un jeune orang (*Pongo pygmaeus* Hoppius) né à la ménagerie du Muséum." [Note Concerning a Young Orang (*Pongo pygmaeus* Hoppius) Born in the Menagerie of the Museum.] *Bulletin du Museum* 3 (1931): 475–78.
Bresard, Bernadette. "Problème de l'assymétrie cérébrale chez les anthropoïdes. Contribution à l'étude de la phylogenèse des processus cognitifs." [The Problem of Cerebral Asymmetry in Anthropoids: Contribution to the Phylogenetic Study of Cognitive Processes.] PhD diss., Université Pierre and Marie Curie (Paris 6), 1983.
Bresson, François. "Inferences from Animal to Man: Identifying Functions." In *Methods of Inference from Animal to Human Behaviour*, edited by Mario von Cranach, 319–42. Paris: Mouthon, 1976.
Burkhardt, Richard W. "La Ménagerie et la vie du Muséum." [The Menagerie and the Life of the Museum.] In *Le Muséum au premier siècle de son histoire* [*The Museum during the First Century of Its History*], edited by Claude Blanckaert, Claudine Cohen, Pietro Corsi, and Jean-Louis Fischer, 481–508. Paris: Éditions du Muséum national d'Histoire naturelle, 1997.
Bussolini, Jeffrey. "What Is a Dispositive?" *Foucault Studies* 10 (2010): 85–107.

Byrne, Richard R., and Anne Russon. "Learning by Imitation: A Hierarchical Approach." *Behavioral and Brain Sciences* 21 (1998): 667–721.

Chamove, Arnold S., Geoffrey R. Hosey, and Peter Schaetzel. "Visitors Excite Primates in Zoos." *Zoo Biology* 7 (1988): 359–69.

Cuvier, Georges, and Etienne Geoffroy Saint-Hilaire. "Histoire naturelle des orangs-outangs." [Natural History of the Orangutans.] *Magasin Encyclopédique* 1 (1795): 451–63.

Cuvier, Georges, and Etienne Geoffroy Saint-Hilaire. "Mémoire sur les orangs-outangs." [Memoir on Orangutans.] *Journal de Physique, de Chimie et d'Histoire naturelle* 3 (1798): 185–91.

Deleuze, Gilles. "A for Animal." In *L'abécédaire de Gilles Deleuze,* with Claire Parnet, filmed by Pierre-André Boutang. Production Sodaperaga, 1988–89.

Demars, Christian, and Michel Goustard. "Structure et règles de déroulement des émissions sonores des Hylobates, *Hylobates concolor.*" [Structure and rules of sound emission in crested gibbons (*Hylobates concolor*).] *Bulletin Biologique de la France et de la Belgique* 106 (1973): 177–91.

De Waal, Frans. "Bonobos, Left and Right: Primate Politics Heats Up Again as Liberals and Conservatives Spindoctor Science." *Skeptic,* August 8, 2007. http://www.skeptic.com/eskeptic/07-08-08/

Egenter, Nold. "Evolutionary Architecture: The Nest Building Behavior of Higher Apes." *International Semiotic Spectrum* 14 (1990).

Endler, John A., Lorna C. Endler, and Natalie R. Doerr. "Great Bowerbirds Create Theaters with Forced Perspective When Seen by Their Audience." *Current Biology* 20 (2010): 1679–84.

Estebanez, Jean. "Le zoo comme dispositif spatial: mise en scène du monde et de la juste distance entre l'humain et l'animal." [The Zoo as a Spatial Structure: Staging the World and the Appropriate Distance between Humans and Animals.] *L'Espace Géographique* 39 (2010): 172–79.

Fernandez-Carriba, Samuel. "Los chimpancés que trituran la comida: un ejemplo de transformación del alimento en primates no humanos." [Chimpanzees That Crush Their Food: An Example of Food Processing in Nonhuman Primates.] *Boletín de la Asociación Primatológica Española* 7 (2000): 7–8.

Fernandez-Carriba, Samuel, and Angela Loeches. "Fruit Smearing by Captive Chimpanzees: A Newly Observed Food Processing Behavior." *Current Anthropology* 42 (2001): 143–47.

Foucault, Michel. "The Confession of the Flesh." Interview. In *Power/Knowledge: Selected Interviews and Other Writings, 1972–1977,* edited by Michel Foucault, translated by Colin Gordon, 194–228. New York: Pantheon Books, 1980.

Fouts, Roger S., Alan D. Hirsch, and Deborah H. Fouts. "Cultural Transmission of a Human Language in a Chimpanzee Mother-Infant Relationship." In *Child Nurturance: Studies of Development in Nonhuman Primates,* edited by Hiram E. Fitzgerald, John A. Mullins, and Patricia Gage, 159–93. New York: Plenum, 1982.

Fruth, Barbara, and Gottfried Hohmann. "Comparative Analyses of Nest Building Behavior in Bonobos and Chimpanzees." In *Chimpanzee Cultures,* edited by Richard W. Wrangham, W. C. McGrew, Frans B. M. de Waal, and Paul G. Heltne, 109–28. Cambridge, MA: Harvard University Press, 1996.

Fruth, Barbara, and Gottfried Hohmann. "Nests: Living Artefacts of Recent Apes?" *Current Anthropology* 35 (1994): 310–11.

Galdikas, Biruté M. F. "Orang-utan Tool Use at Tanjung Putting Reserve, Central

Sudonesian Borneo (Kalimantan Tengah)." *Journal of Human Evolution* 10 (1982): 19–33.

Gewalt, W. "Orang-Utans (*Pongo pygmaeus*) als 'Seiler.'" *Zeitschrift für Säugetierkunde* 40 (1975): 320–21.

Gilliard, Thomas E. "Bower Ornamentation versus Plumage Characters in Bower-Birds." *Auk* 73, no. 3 (1956): 450–51.

Goodall, Jane M. "Nest-Building Behavior in the Free Ranging Chimpanzee." *Annals of the New York Academy of Sciences* 102 (1962): 455–67.

Gould, Stephen J., and Elisabeth S. Vrba. "Exaptation—A Missing Term in the Science of Form." *Paleobiology* 8 (1982): 4–15.

Groves, Colin J., and Jordi Sabater Pi. "From Ape's Nest to Human Fix-point." *Man* 20 (1985): 22–47.

Hall-Craggs, Joan. "The Aesthetic Content of Bird Song." In *Bird Vocalizations*, edited by Robert A. Hinde, 367–81. Cambridge: Cambridge University Press, 1969.

Hamy, Ernest-Théodore. "Les derniers jours du Jardin du Roi et la fondation du Muséum d'histoire naturelle." [The Last Days of the Jardin du Roi and the Foundation of the Museum of Natural History.] In *Centenaire de la fondation du Muséum d'Histoire naturelle, 10 juin 1793–10 juin 1893, volume commémoratif publié par les Professeurs du Muséum.* [*Centenary of the Foundation of the Natural History Museum, 10th June 1793–10th June 1893. Commemorative volume published by the professors of the museum.*] Paris: Imprimerie Nationale, 1893.

Haraway, Donna. "Teddy Bear Patriarchy: Taxidermy in the Garden of Eden, New York City, 1908–1936." *Social Text* 11 (1984): 19–64.

Hediger, Heini. "Un problème qui nous ramène à l'homme: l'habitat des animaux." [A Problem That Brings Us Back to Humans: The Habitats of Animals.] In *La pensée non humaine* [*Nonhuman Thought*], edited by Jacques Graven, 219–25. Encyclopédie Planète. Paris: Planète, 1963.

Hediger, Heini. "Nest and Home." *Folia Primatologica* 28 (1977): 170–87.

Herzfeld, Chris. "L'invention du bonobo." [The Invention of the Bonobo.] *Bulletin d'Histoire et d'Épistémologie des Sciences de la Vie* 14 (2007): 139–62.

Herzfeld, Chris. "Pourquoi les nœuds des grands singes nous intéressent-ils tellement?" [Why Are We So Interested in the Knots of Great Apes?] (catalogue de l'exposition Bêtes et Hommes, Parc de la Villette). Paris: Gallimard, 2007.

Herzfeld, Chris, and Dominique Lestel. "Knot Tying in Great Apes: Etho-ethnology of an Unusual Tool Behavior." *Social Science Information* 44 (2005): 621–53.

Herzog, George. "Do Animals Have Music?" *Bulletin of the American Musicological Society* 5 (1941): 3–4.

Jordan, Claudia. "Object Manipulation and Tool Use in Captive Pygmy Chimpanzees (*Pan paniscus*)." *Journal of Human Evolution* 11 (1982): 35–39.

Kelley, Laura A., and John A. Endler. "Illusions Promote Mating Success in Great Bowerbirds." *Science* 335 (2012): 335–38.

Kortlandt, Adriaan, and M. Kooij. "Protohominid Behavior in Primates (Preliminary Communication)." *Symposia of the Zoological Society of London* 10 (1963): 61–88.

Kortlandt, Adriaan. "Handgebrauch bei freilebenden Schimpansen." [Hand Use in Free-Living Chimpanzees.] In *Handgebrauch und Verständigung bei Affen und Frühmenschen* [*Hand Use and Communication in Apes and Early Humans*], edited by Bernhard Rensch, 59–102. Bern: Huber, 1968.

Krützen, Michael, Erik P. Willems, and Carel P. van Schaik. "Culture and Geographic Variation in Orangutan Behavior." *Current Biology* 21 (2011): 1808–12.

Lacour, Marie-Christine. "Communications interspécifiques (Chimpanzé, orang outan, homme). Exploitation d'un recueil de séquences d'expressions faciales : Préparation à l'informatisation." [Interspecific Communication (Chimpanzee, Orangutan, Human). Exploitation of Recordings of Sequences of Facial Expression: Preparation for Computerization.] Master's thesis, Université Paris Diderot (Paris 7), 1983.

Langaney, André. "Watana, orang-outan immigrée." [Watana, an immigrant orangutan.] *Le Temps (Genève)*, June 15, 2004.

Lestel, Dominique, and Chris Herzfeld. "L'intelligence des grands singes en question." [The Intelligence of Great Apes Examined.] *Le Courrier de la Nature (Société Nationale de Protection de la Nature), Numéro spécial Grands Singes* 227 (2006): 28–33.

Lestel, Dominique, and Chris Herzfeld. "Topological Ape: Knots-Tying and Untying and the Origins of Mathematics." In *Images and Reasoning, Interdisciplinary Conference Series on Reasoning Studies*, Vol. 1, edited by Pierre Grialou, Giuseppe Longo, and Mitsuhiro Okada, 147–62. Tokyo: Keio University Press, 2005.

Lethmate, Jürgen. "Instrumental Behaviour of Zoo Orang-utans." *Journal of Human Evolution* 8 (1979): 741–44.

Lethmate, Jürgen. "Nestbauverhalten eines isoliert aufgezogenen jungen Orang-Utans." [Nest-Building Behavior of an Isolated Hand-reared Young Orangutan.] *Primates* 18 (1977): 545–54.

Lethmate, Jürgen. "Tool-using Skills of Orang-utans." *Journal of Human Evolution* 11 (1982): 49–64.

Lisman, John E., and Anthony A. Grace. "The Hippocampal-VTA Loop Controlling Entry into Long-Term Memory." *Neuron* 46 (2005): 703–13.

Locke, Devin P., LaDeana W. Hillier, Wesley C. Warren, et al. "Comparative and Demographic Analysis of Orang-utan Genomes." *Nature* 469 (2011): 529–33.

Lundgren, Thomas S., and Nagi N. Mansour. "Vortex Ring Bubbles." *Journal of Fluid Mechanics* 224 (1991): 177–96.

Mâche, François-Bernard. "Les oiseaux musiciens." [Bird Musicians.] *Sciences et Avenir, Hors-Série Paroles Animales* 131 (2002): 62–68.

Madden, Joah R., Caroline Dingle, Jess Isden, Janka Sparfeld, Anne W. Goldizen, and John A. Endler. "Male Spotted Bowerbirds Propagate Fruit for Use in Sexual Display." *Current Biology* 22 (2012): R264–65.

Marten, Ken, Karim Shariff, Suchi Psarakos, and Don White. "Les ronds d'air des dauphins." [The Air Rings of Dolphins.] *Pour la Science* 228 (1996): 90–95.

Martinelli, Dario. "How Musical Is a Whale? Towards a Theory of Zoömusicology." In *Acta Semiotica Fennica: Approaches to Musical Semiotics*, edited by E. Tarasti, Hakapaino, International Semiotics Institute. Helsinki: Semiotic Society of Finland, 2002.

Massumi, Brian. "Ceci n'est pas une morsure: Animalité et abstraction chez Deleuze et Guattari." [This Is Not a Bite: Animality and Abstraction According to Deleuze and Guattari.] *Philosophie* 112 (2011): 67–91.

Maxman, Amy. "Dopamine's Role Linked to Emotions." *Science News*, August 2, 2008.

McCowan, Brenda, Lori Marino, Erik Vance, Leah Walke, and Diana Reiss. "Bubble Ring Play of Bottlenose Dolphins (*Tursiops truncatus*): Implications for Cognition." *Journal of Comparative Psychology* 114 (2000): 98–106.

McGrew, William C., and Linda F. Marchant. "Chimpanzee Wears a Knotted Skin 'Necklace.'" *Pan Africa News* 5 (1998): 8–9.

Miles, Lyn H., Robert W. Mitchell, and Stephen E. Harper. "Simon Says: The

Development of Imitation in an Enculturated Orangutan." In *Reaching into Thought: The Minds of the Great Apes*, edited by Anne E. Russon, Kim A. Bard, and Sue Taylor Parker, 278–99. Cambridge: Cambridge University Press, 1996.

Montane, Louis. "A Cuban Chimpanzee." *Journal of Animal Behavior* 6 (1916): 330–33.

Panksepp, Jaak. "Affective Consciousness: Core Emotional Feelings in Animal and Humans." *Consciousness and Cognition* 14 (2005): 30–80.

Panksepp, Jaak. "On the Embodied Neural Nature of Core Emotional Affects." *Journal of Consciousness Studies* 12 (2005): 158–84.

Parker, Ian. "Swingers: Bonobos Are Celebrated as Peace-Loving, Matriarchal, and Sexually Liberated. Are They?" *New Yorker*, July 30, 2007.

Polster, Burkard. "What Is the Best Way to Lace Your Shoes?" *Nature* 420 (2002): 476.

Porcher, Jocelyne. "Le travail dans l'élevage industriel des porcs. Souffrance des animaux, souffrance des hommes." [Working in Industrial Pig Farming: Suffering for Animals, Suffering for People.] In *Les animaux d'élevage ont-ils droit au bien-être? [Have Animals on Farms the Right to Well-Being?]*, edited by Florence Burgat and Robert Dantzer, 25–64. Paris: INRA, 2001.

Pradhan, Gauri R., Maria A. van Noordwijk, and Carel van Schaik. "A Model for the Evolution of Developmental Arrest in Male Orangutans." *American Journal of Physical Anthropology* 149 (2012): 18–25.

Raven, Henry C. "Further Adventures of Meshie. A Chimpanzee That Has Lived Most of Her Life in a New York Suburban Home." *Natural History* 33 (1933): 607. (Reprinted: *Natural History*, July/August 2002.)

Richard, Bernard. "Les mammifères constructeurs." [Mammals that Construct.] In *La Recherche en éthologie. Les comportements animaux et humains [Research in Ethology: Animal and Human Behavior]*, edited by Jean-Pierre Desportes and Assomption Vloebergh, 156–73. Collection Points Sciences/La Recherche. Paris: Éditions du Seuil, 1979.

Rumbaugh, Duane M., and Timothy V. Gill. "The Learning Skills of Great Apes." *Journal of Human Evolution* 2 (1973): 171–72.

Russon, Anne E., and Birutė M. Galdikas. "Imitation in Free-Ranging Rehabilitant Orangutans (*Pongo pygmaeus*)." *Journal of Comparative Psychology* 107 (1993): 147–61.

Sabater Pi, Jordi, Joaquim J. Vea, and Jordi Serrallonga. "Did the First Hominids Build Nests?" *Current Anthropology* 38 (1997): 914–16.

Strum, Shirley, and Bruno Latour. "Redéfinir le lien social: des babouins aux humains." [Redefining the Social bond: From Baboons to Humans.] In *Sociologie de la traduction, Textes fondateurs [Sociology of Translation: Founding Texts]*, edited by Madeleine Akrich and Michel Callon, 71–86. Paris: Presses de l'École des mines de Paris, 2006.

Szöke, Peter. "Ornitomuzikologia." [Ornithomusicology.] *Magyar Tudomany* 9 (1963): 592–607.

Tratz, Eduard P., and Heinz Heck. "Der afrikanische Anthropoide "Bonobo": Eine neue Menschenaffengattung." [The African Ape "Bonobo": A New Great Ape Genus.] *Säugetierkundliche Mitteilungen* 2 (1954): 97–101.

Van Elsacker, Linda., and Vera Walraven. "The Spontaneous Use of a Pineapple as a Recipient by a Captive Bonobo (*Pan paniscus*)." *Mammalia* 58 (1994): 159–62.

Van Schaik, Carel P., Marc Ancrenaz, Gwendolyn Borgen, et al. "Orangutan Cultures and the Evolution of Material Culture." *Science* 299 (2003): 102–5.

Vauclair, Jacques. "Would Humans without Language Be Apes?" In *Cultural Guidance in the Development of the Human Mind,* Vol. 7, *Advances in Child Development within Culturally Structured Environments,* edited by J. Valsiner and Aaro Toomela, 9–26. Greenwich, CT: Ablex Publishing Corporation, 2003.

Whiten, Andrew, Jane Goodall, William C. McGrew, et al. "Cultures in Chimpanzees." *Nature* 399 (1999): 682–85.